Recent Trends in Traditional and Technical Textiles

Vinay Midha · A. Mukhopadhyay
Editors

Recent Trends in Traditional and Technical Textiles

Select Proceedings of ICETT 2019

Springer

Editors
Vinay Midha
Department of Textile Technology
Dr. B. R. Ambedkar National Institute of Technology
Jalandhar, Punjab, India

A. Mukhopadhyay
Department of Textile Technology
Dr. B. R. Ambedkar National Institute of Technology
Jalandhar, Punjab, India

ISBN 978-981-15-9994-1 ISBN 978-981-15-9995-8 (eBook)
https://doi.org/10.1007/978-981-15-9995-8

This Springer imprint is published by the registered company Springer Nature Singapore Pte Ltd.
The registered company address is: 152 Beach Road, #21-01/04 Gateway East, Singapore 189721, Singapore

Contents

About the Editors

Dr. Vinay Midha is a Professor in the Department of Textile Technology at Dr. B. R. Ambedkar National Institute of Technology, Jalandhar, India. His academic and research experience spans over 25 years after having one year industrial experience during which he worked on several research projects supported by various funding agencies. His specialization and areas of interest include sewing threads, dynamics of sewing process, non-woven fabrics, waterproof breathable fabrics, medical and healthcare textiles, and runoff erosion control using natural fibre geomeshes. He has contributed over 100 research papers in reputed international and national peer reviewed journals and international conferences. He has written seven book chapters for Woodhead Publications and has also filed two patents. He is also involved in the review process for many reputed journals in the field of textile technology.

Dr. A. Mukhopadhyay is a Professor in the Department of Textile Technology at National Institute of Technology, Jalandhar, India. His academic and research experience spans over 28 years with two years of industrial experience. His field of interest is diverse and interdisciplinary in nature encompassing filtration science and technology especially in industrial gas filtration and pollution control. He has contributed to more than 180 research publications in reputed academic and applied journals, conference proceedings etc. His publications also include 12 monographs/book chapters encompassing different areas of technical textiles. He has successfully carried out several consultancy and research project supported by DST, CSIR, MHRD, TIFAC and World Bank Assisted Project. He presented his research at international conferences held in countries like the USA, UK, Germany, Austria, Slovenia, South Korea, and China. He has been a scientific committee member of International Nonwoven Technical Conferences (INTC 2007 & INTC 2008 held in the USA), World Filtration Congress (WFC 11, Graz; WFC12, Taiwan, WFC13, San Diego, CA), and also FILTECH (2013-2019 held in Germany). He also chaired sessions at international conferences such as World Filtration Congress WFC11 held at Graz, Austria on April 16-20, 2012 and FILTECH Conferences held in Germany (2011-2019).

Influence of Weft Density on Runoff Erosion Control Performance of Rolled Erosion Control Systems

Vinay Kumar Midha and S. Suresh Kumar

1 Introduction

Landscapes disturbed by human activities commonly have erosion rates accelerated by several orders of magnitude over pre-disturbance conditions. New approaches to effectively decrease soil erosion rates from disturbed lands are essential and adopted in the recent past to protect major environmental damages. One such best management practice (BMP) combines the application of rolled erosion control systems (RECSs), composed of either natural or synthetic fibers, with seeding of hill slopes to enhance biomass production. This synergistic approach is one of the most appropriate practices for mitigating excessive soil erosion on disturbed non-agricultural hill slopes [1–4]. However, RECSs effectively mitigate runoff erosion, but limited studies are reported yet on the characteristics of RECSs that are beneficial for the potential reduction in erosion or enhancement of grass production (Vegetation growth) [5–8]. Commercial RECSs are not engineered or tailored products for onsite conditions. The various designs of commercial RECSs that are marketed today by the Erosion control Industries have probably been developed through a combination of trial-and-error and some knowledge of those properties that may influence water erosion processes. In a critical review paper, it is suggested that without a more detailed understanding of the influence of various physical attributes of RECSs on soil erosion processes and production of vegetation, there is little chance that major advances can be made in the future design of RECSs [9, 10].

To understand the physical attributes of RCESs, it is necessary to understand its structural construction. Commercial biodegradable RECSs are generally woven structures with two set of yarns, namely vertical "warp yarns" and horizontal "weft

V. K. Midha (✉) · S. S. Kumar
Department of Textile Technology, Dr. B. R. Ambedkar National Institute of Technology, Jalandhar, Punjab 144011, India
e-mail: midhav@rediffmail.com

V. Midha and A. Mukhopadhyay (eds.), *Recent Trends in Traditional and Technical Textiles*, https://doi.org/10.1007/978-981-15-9995-8_1

yarns." During onsite application, RECSs are rolled down the slope in such a way that the vertical warp yarns lies toward the direction of slope, whereas the horizontal weft yarn lies against the direction of slope [11–14]. At this condition, the weft yarn lays against the direction of slope reduces the velocity of surface runoff by creating more number of micro-barriers due its three-dimensional structure. Hence, the variation in physical and structural construction of cross-laid weft yarn (yarns/meter) would improve the erosion control performance of RECSs. Based on onsite laying conditions (ideal conditions), a geometrical model is proposed to calculate the storage volume of RECSs to evaluate its performance [7, 8]. In the proposed model, Sanyal considered the diameter of weft yarns of RECSs as circular cross section, then the hindrance (micro-barriers) created by the weft yarns against the slope was expected to be a triangular storage effect. In accordance with his model, Sanyal increased the diameter of jute RECSs and achieved an improved erosion control performance. However, increase in diameter of yarns leads to increase in flexural rigidity of RECS, which reduces the contact of the RECS with the soil, and therefore may reduce the erosion control performance. Increase in weft yarn density (weft yarns per meter) of RECSs can also increase the storage volume and hindrance against the overland flow and is expected to improve the performance of RECSs (Fig. 1).

But till date no such study has been reported regarding such structural modification of RECSs. Therefore, in this study 15 different RECSs (Coir and Jute) with different weft density (weft yarns per meter) are tested for erosion control performance at different slope angles (i.e., 15°, 30° and 45°) and germination performance.

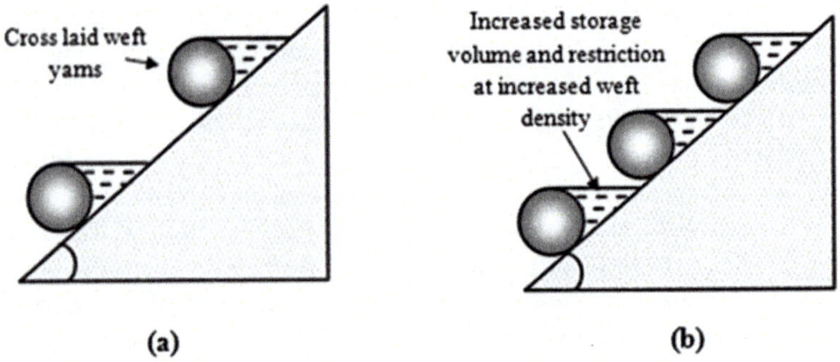

Fig. 1 Storage effect and restriction created by cross-laid weft yarns of RECs at different weft density: **a** minimum weft density; **b** increased weft density

2 Materials and Methodology

2.1 *Materials*

Commercial coir and jute yarns with different linear density are selected and woven to prepare 15 different RECSs samples with different weft density (Table 1). At increased weft density, weft per meter increases and the distance between the successive weft yarns decreases. So, weft density for different yarn counts is maintained constant by maintaining distance between the adjacent weft yarns. In case of coir RECSs distance of 17, 19, and 21 mm are maintained between adjacent weft yarns, to weave different RECSs of different weft density. Whereas in jute RECSs distance of 8, 12, and 16 mm is maintained between adjacent weft yarns. Flexural rigidity, an important property of RECSs that influences its erosion control performance, is expected to change as the weft yarn density changes. Therefore, flexural rigidity of all RECSs is measured according to ASTM D1388 (wider width of 25 cm) at dry and wet conditions [9].

2.2 *Runoff Erosion Test*

To perform runoff erosion control test, RECS samples are subjected to simulated rainfall conditions in a runoff erosion control tester (bench-scale), which is designed based on ASTM D 7101 standards with certain modification in the ramp size (Fig. 2a) [15, 16].

Since RECSs are open weave structures with larger mesh opening size (19 X 19 cm and 10 X 12 cm), the test core of 20 cm diameter used in ASTM D7101 is substituted with ramp size of 50 cm length and 25 cm width [17]. V-jet nozzles are used in the rainfall simulators to simulate rainfall intensity 100 mm/hr. Runoff erosion test for different RECSs samples is carried out at three different slope angles of 15°, 30°, and 45° by changing the slope angle of ramp in the runoff erosion setup (Fig. 2b, c) carried with soil (soil infiltration condition) and without soil (zero-infiltration condition) conditions.

2.2.1 Runoff Test at Soil Infiltration Condition

In soil infiltration condition, soil from Shivalik/lower Himalayan regions (31° 37′ 48.2″ N latitude 76° 00′ 47.5″ E longitude) is filled in test core/soil tray of dimension 50 cm length, 25 cm width, and 25 cm depth [18, 19] (Fig. 2b). Based on ASTM D 698 standards soil is compacted in test tray at 15% moisture content by dropping 5.50-lbf (24.5-N) rammer from the height of 30 cm [20]. Prepared soil trays are positioned at different slope angles on the runoff erosion setup (covered with RECS) and subjected to required rainfall intensity for 3 min in each trial and

Table 1 Specifications of coir and jute RECSs

Material	Sample code	Linear density of warp and weft yarn (Tex)	Diameter of warp and weft yarn (mm)	Yarn density (ends/m)	Yarn density (picks/m) PPM	Mesh opening Size (mm)	Distance between adjacent weft yarns (mm)	Weight per unit area (g/m^2)	Fractional cover	Flexural rigidity (μNm)			
										Dry		Wet	
										Warp	Weft	Warp	Weft
Coir	C1	4794	4	43	40	19 × 21	21	397.90	0.30	4451	4270	1654	1503
	C2	4794	4	43	43	19 × 19	19	412.28	0.31	4380	4380	1581	1581
	C3	4794	4	43	48	19 × 17	17	436.25	0.33	4268	4429	1498	1624
	C4	6765	5	42	38	19 × 21	21	541.20	0.36	5842	5641	2054	1872
	C5	6765	5	42	42	19 × 19	19	568.26	0.38	5759	5759	1987	1987
	C6	6765	5	42	45	19 × 17	17	588.56	0.39	5610	5799	1903	2068
	C7	7614	6.5	39	36	19 × 21	21	571.05	0.43	6621	6452	2584	2451
	C8	7614	6.5	39	39	19 × 19	19	593.89	0.44	6501	6501	2492	2492
	C9	7614	6.5	39	43	19 × 17	17	624.35	0.46	6403	6608	2431	2561
Jute	J1	2961	2.5	80	54	10 × 16	16	396.93	0.31	923	712	232	56
	J2	2961	2.5	80	69	10 × 12	12	441.09	0.34	861	668	189	72
	J3	2961	2.5	80	95	10 × 8	8	518.88	0.39	742	509	151	84
	J4	4773	3.9	72	50	10 × 16	16	583.51	0.42	1021	804	243	59
	J5	4773	3.9	72	63	10 × 12	12	643.84	0.46	972	718	196	81
	J6	4773	3.9	72	84	10 × 8	8	744.75	0.52	954	792	156	108

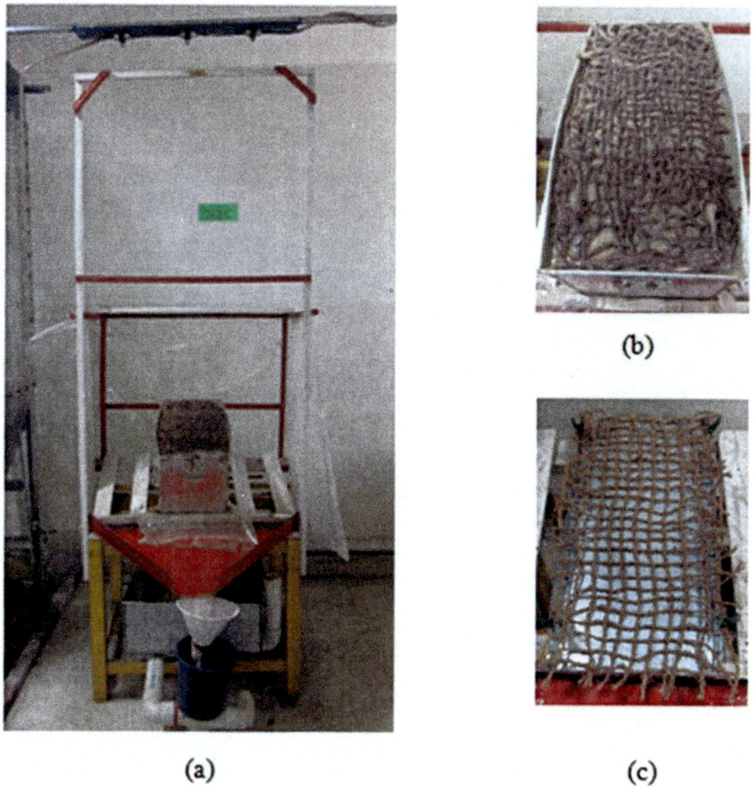

Fig. 2 Runoff erosion test at different infiltration condition: **a** runoff erosion setup; **b** soil infiltration condition; **c** zero-infiltration

runoff water with eroded soil is collected. Eighteen such trials are repeated for each sample, and results of initial three trials are not considered in those results to avoid initial error that may be generated due to the initial absorption of water by the soil and RECSs. Similarly testing is performed in uncovered soil trays (without RECSs), to evaluate control test performance. The same procedure is repeated for different samples at different slope angles; however, newly prepared test trays are used for each sample, slope angle, and rainfall intensity. The eroded soil from the runoff water is separated by sedimentation process, and the erosion control % of each sample is calculated using Eq. 1 [15, 16].

$$\text{Erosion control } \% = \frac{(E - C)}{E} \times 100 \tag{1}$$

Where,

E Eroded soil without RECSs (gms).
C Eroded soil with RECSs (gms).

2.2.2 Runoff Test at Zero-Infiltration Condition

Since, performing runoff erosion test at soil infiltration conditions is influenced by the water restriction behavior of RECSs, due to the absorption of water by the soil, and the runoff erosion test is also performed at zero-infiltration (without soil) condition for better understanding of the influence of weft yarn diameter on the reduction in runoff velocity. To perform runoff erosion test at zero-infiltration condition, soil tray is replaced with a metallic ramp with uniform smooth surface (Fig. 2c) [17]. RECS sample is laid over the metallic ramp and subjected to simulated rainfall for 3 min to collect runoff water volume. Subsequently rainfall is ceased for 3 min to collect the culmination discharge. Similar procedure is repeated for evaluating the performance of different RECSs at different slope angles.

2.3 Germination Test

Germination test is performed according to ASTM D 7322 by using earthen pots filled with site soil, and sown with equal number of wheat seeds (60 seeds/pot). The pots are covered with RECSs at surface of the soil as shown in the Fig. 3a, and 1 pot is left uncovered for the control test. The test pots are kept at uniform temperature and lighting conditions, and watered until the 21st day. At the end of the 21st day, percentage of vegetation in each pot is calculated using Eq. 2 with reference to the number of plants germinated in the control test pot.

$$\%\text{ of vegetation} = \left(\frac{\text{Number of plants germinated in the pot with RECS}}{\text{Number of plants germinated in the pot without RECS}}\right) \times 100 \tag{2}$$

Other important factors that affect the soil stability and germination of plants are the number and length of roots; so total rooting is also calculated at the end of 21st

(a) (b)

Fig. 3 Germination test: **a** germination earthen pots; **b** uprooted vegetation at 21st day for evaluation of rooting day. Primary roots in wheat plants germinate in 10 days and secondary roots start appearing after 20 days [21–23]. After 21 days, the average primary root length and average number of roots/plants are measured in each pot by uprooting 10 plants randomly (Fig. 3b). Total rooting after 21 days is calculated using Eq. 3 [15, 16].

$$\text{Total rooting(cm)} = N \times n \times L \tag{3}$$

where,

N Total number of plants germinated after 21 days,
n Average number of primary roots/plants,
L Average primary root length (cm).

3 Result and Discussion

3.1 Influence of Weft Density on Erosion Control Performance

Erosion control performance of coir and jute RECSs at different weft density, slope angle, and infiltration is tabulated in Table 2.

3.1.1 Erosion Control Performance at Soil Infiltration Condition

At soil infiltration condition, coir RECSs with finer yarns (4794 and 6765 Tex) show lower runoff volume and improved erosion control percentage at increased weft yarn density (lower distance between weft yarns), at all slope angles (15°, 30° and 45°) (Table 2 and Fig. 4).

It is due to the higher storage volume and more restriction offered by higher number of weft yarns against the runoff water running down the slope [12, 13]. However, in coarser yarns (7614 Tex) RECSs, increased weft density results in initial increase in erosion control percentage, but further increase in weft density resulted in reduced erosion control percentage at steeper angle of slope (45°). It is due the increased flexural rigidity of coir RECSs (Table 1) and higher rate of erosion at steeper slope. At steeper slopes "gully erosion" with deeper contours would be formed [24]. It is difficult for the high flexural rigidity coir RECSs with coarser yarns (7614 in both warp and weft direction) to drape (Fig. 5) properly over these contours. This reduces the contact between soil surface and RECSs, and results in reduced erosion control performance at steeper slopes.

At soil infiltration condition, Jute RECSs with higher weft density (minimum distance between weft yarns) result in reduced runoff volume and improved erosion control performance, at all slope angles (15°, 30° and 45°) (Table 2). This is due to increased storage volume and restriction against runoff water at increased weft

Table 2 Erosion control performance of coir and jute RECSs at soil and zero-infiltration condition

Material	Sample code	Warp and weft yarn linear density (Tex)	Mesh opening size (mm)	Soil infiltration						Zero-infiltration					
				Runoff volume (ml)			Erosion Control %			Runoff volume (ml)			Culmination discharge (ml)		
				15°	30°	45°	15°	30°	45°	15°	30°	45°	15°	30°	45°
Coir	C1	4794	19 × 21	423	441	472	44.2	38.2	34.2	552	574	589	42	52	56
	C2	4794	19 × 19	402	424	454	52.3	42.9	38.7	543	542	575	51	59	64
	C3	4794	19 × 17	385	397	429	57.2	48.7	43.5	518	531	568	62	68	71
	C4	6765	19 × 21	401	417	443	58.2	49.7	46.3	567	578	594	47	59	69
	C5	6765	19 × 19	378	408	430	61.7	53.3	50.2	560	558	571	56	68	62
	C6	6765	19 × 17	362	383	418	63.2	56.9	52.9	544	529	541	65	74	78
	C7	7614	19 × 21	377	391	449	62.7	60.1	36.9	563	585	572	51	56	65
	C8	7614	19 × 19	360	376	441	64.9	62.4	48.5	554	580	564	49	58	73
	C9	7614	19 × 17	342	354	428	66.2	56.4	37.2	550	578	589	52	59	69
Jute	J1	2961	10 × 16	361	385	402	71.8	67.3	64.2	402	422	447	162	173	185
	J2	2961	10 × 12	345	367	388	76.2	71.5	68.5	379	410	435	171	180	206
	J3	2961	10 × 8	318	336	369	81.3	77.8	72.3	354	381	412	183	194	228
	J4	4773	10 × 16	348	369	378	72.8	69.7	67.1	372	389	414	162	178	195
	J5	4773	10 × 12	309	322	341	79.1	74.6	71.3	358	375	390	209	220	228
	J6	4773	10 × 8	262	294	327	85.9	79.4	74.3	331	347	362	225	237	242

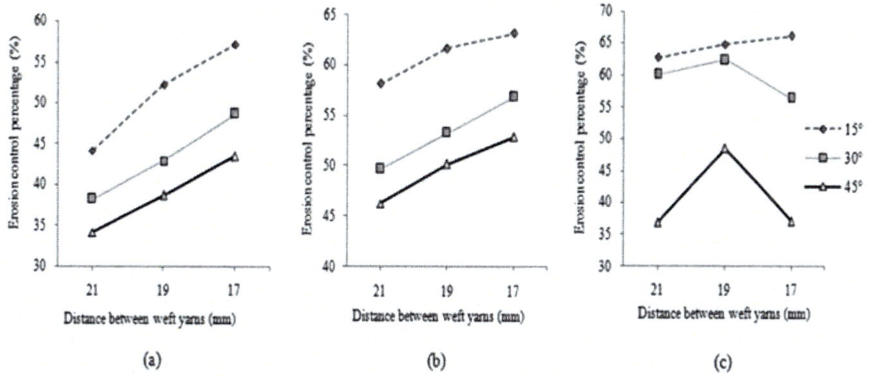

Fig. 4 Erosion control percentage of coir RECs at different weft density and slope angle: **a** yarn count of 4794 Tex; **b** yarn count of 6765 Tex; **c** yarn count of 7614 Tex

Fig. 5 Drapability of coir RECSs at different slope angle: **a** sheet erosion and lower slope angle; **b** gully erosion at steeper slope angle

density, and also due to the better drapability of jute RECSs. Even at increased weft density, Jute RECSs has lower flexural rigidity as compared to the coir RECSs (Table 1), due to finer and softer jute fibers. Further, flexural rigidity of jute RECSs reduces significantly under wet conditions, as compared to coir RECSs [25, 26]. Weft directional flexural rigidity of jute RECS J1 is 712 μNm and 56 μNm under dry and wet conditions, respectively. It is due to higher water absorbing capacity and finer fibers of jute RECSs, which facilitate better contact between jute RECSs and soil surface and results in better erosion control performance as compared to coir RECSs.

Influence of weft density on erosion control percentage of coir and jute RECSs is statistically analyzed using Analysis of Variance (ANOVA). In coir RECSs, it is observed that the slope angle has the maximum influence on erosion control percentage with a contribution of 47.29%, and it is followed by the yarn count and weft density with a contribution of 27.39% and 7.28%, respectively (Table 3).

Table 3 ANOVA analysis of erosion control percentage of coir RECSs

Type of RECSs	Source	DF	Seq ss	Contribution %	Adj SS	ADj MS	F-value	P-value
Coir RECSs	Yarn linear density (Tex)	2	654.29	27.39	654.29	327.145	63.60	0.000
	Weft density (distance between weft yarn) (mm)	2	173.98	7.28	173.98	86.991	16.91	0.000
	Slope angle (degree)	2	1129.51	47.29	1129.51	564.757	109.80	0.000
	Yarn tex * weft density	4	122.23	5.12	122.23	30.556	5.94	0.007
	Yarn tex * slope angle (degree)	4	246.86	10.33	246.86	61.715	12.00	0.000
	Error	12	61.72	2.58	61.72	5.144		
	Total	26	2388.60	100.00				
Jute RECSs	Yarn linear density (tex)	1	30.161	5.72	30.161	30.161	34.65	0.004
	Weft density (distance between weft yarn) (mm)	2	281.363	53.37	281.363	140.682	161.60	0.000
	Slope angle (degree)	2	203.853	38.67	203.853	101.927	117.08	0.000
	Yarn tex * weft density (mm)	2	0.568	0.11	0.568	0.284	0.33	0.739
	Yarn tex * slope angle (degree)	2	0.164	0.03	0.164	0.082	0.09	0.912
	Weft density (mm) * slope angle (degree)	4	7.553	1.43	7.553	1.888	2.17	0.236
	Error	4	3.482	0.66	3.482	0.871		
	Total	17	527.145	100.00				

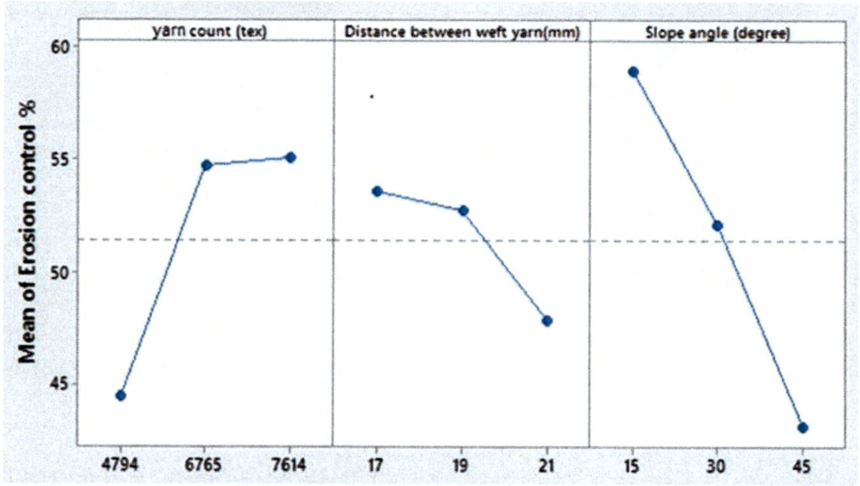

Fig. 6 Main effect plot of erosion control % (ANOVA analysis of coir RECSs)

From the main effects plot (Fig. 6), it is observed that increase in weft density (i.e., lower distance between weft yarns), increases the erosion control percentage of RECSs, whereas increase in slope angle reduces its erosion control percentage. Also, it is observed that the increase in yarn linear density (Tex) initially increases the erosion control percentage significantly, but further increase in yarn count results in marginal increase in erosion control percentage.

In jute RECSs, it is observed that the increase in weft density has the maximum influence on erosion control percentage with contribution percentage of 53.37%, and it is followed by the slope angle and yarn count with contribution of 38.67% and 5.72%, respectively (Table 3).

3.1.2 Erosion Control Performance at Zero—Infiltration Condition

At zero-infiltration condition, overall runoff volume of all coir RECSs is observed to reduce at increased weft density; however, the rate of reduction in runoff volume at different slope angles is inconsistent. The runoff volume of C3 coir RECS at 30° slope angle is 529 ml, whereas at same condition at 15° slope angle the runoff volume is 544 ml. In general, runoff volume increases at increased slope angle due to faster flow of water down the slope, but the results are inconsistent in coir RECSs at different slope, all though an overall reduction of runoff volume is observed at increased weft density. It is due to the stiffer protruding coir fibers on the surface of coir RECSs. The stiffer protruding fibers reduces the contact between coir RECSs and rigid metallic ramp (Fig. 7) [27–29].

Hence the storage effect and restriction created by the cross laid weft yarns are inconsistent and resulted in inconsistent runoff volume and culmination results.

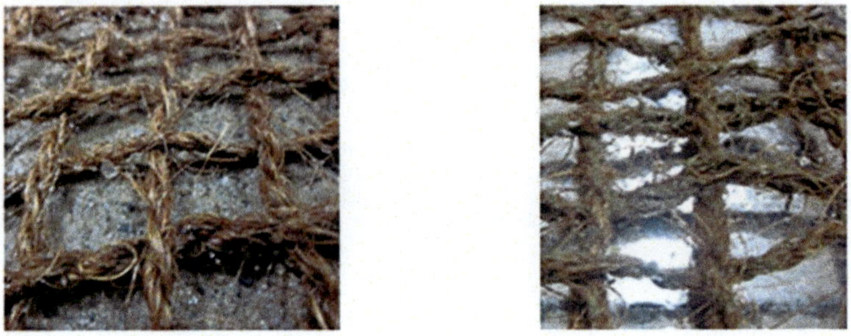

(a) (b)

Fig. 7 Coir RECSs laid at different infiltration condition: **a** soil infiltration condition (soil trays); **b** zero-infiltration condition (metallic ramp)

However, this problem is not observed in soil infiltration condition, because the rigid protruding fibers penetrate into the wet soft soil surface and minimize the problem of reduced contact.

Similar to soil infiltration condition, increase in weft density results in better performance in Jute RECSs at zero-infiltration condition at all slope angles (15°, 30°, and 45°). It is due to the better drapability of jute RECSs [30]. It is observed that the runoff volume and culmination discharge of J1 jute RECS at 30° slope angle is 422 ml and 173 ml, respectively, at zero-infiltration condition. Whereas at same conditions, jute RECS (J3) with higher weft density resulted in lower runoff and higher culmination volume of 381 ml and 194 ml, respectively. This is due to the higher storage volume and higher restriction offered by higher weft density RECSs.

3.2 Influence of Weft Density on Germination Performance

Table 4 shows the germination performance of coir and jute RECSs. It is observed that the increased weft density influences the germination pattern of vegetation due to the variation in fractional cover and flexural rigidity of RECSs.

At increased weft density, fractional cover (surface cover) of RECSs increases (Table 1); this offers more restriction to the growth of vegetation. Hence, at the end of 21st day the percentage of vegetation of coarser yarn coir RECSs (6765 Tex and 7614 Tex) is observed to be lower at increased weft density (Table 4). However, coir RECSs with finer yarn count (4794 Tex) have marginal increase in percentage vegetation, due to its lower flexural rigidity. At lower flexural rigidity, RECSs offer lower restriction to the growth of vegetation even when the fractional cover is high, because the yarns are flexible. Similar trend is observed in jute RECSs, having lower flexural rigidity, percentage of vegetation of jute RECSs is observed to be higher than the coir RECSs, even at higher fractional cover. It is also observed that the jute RECSs facilitate faster initial growth as compared to coir RECSs due to finer and

Table 4 Germination performance of coir and jute RECs

Material	Sample code	Warp and weft yarn linear density (Tex)	Mesh opening Size (mm)	Average number of plants germinated							Vegetation (%)	Average number of roots	Average length of roots (cm)	Total rooting (cm)
				Day 3	Day 6	Day 9	Day 12	Day 15	Day 18	Day 21				
Coir	Control			0.0	12.3	20.7	26.0	28.0	29.0	29.0	100.0	4.0	14.8	1751.6
	C1	4794	19 × 21	0.0	8.0	20.0	24.3	27.0	27.3	27.3	94.8	4.0	14.5	1595.0
	C2	4794	19 × 19	0.0	8.0	18.7	24.0	26.0	27.0	28.0	96.6	4.0	14.5	1624.0
	C3	4794	19 × 17	0.0	7.0	14.0	22.0	25.7	26.0	28.0	96.6	4.1	14.4	1653.1
	C4	6765	19 × 21	0.0	8.0	19.7	24.0	26.0	27.0	27.0	93.1	4.1	13.2	1461.2
	C5	6765	19 × 19	0.0	7.3	17.0	22.0	24.0	25.0	26.3	90.8	4.3	13.2	1494.5
	C6	6765	19 × 17	0.0	6.3	15.0	18.7	21.0	24.0	26.0	89.7	4.2	13.4	1463.3
	C7	7614	19 × 21	0.0	5.0	10.7	21.0	24.0	24.0	24.7	85.0	5.0	11.6	1430.3
	C8	7614	19 × 19	0.0	6.0	17.0	21.0	22.0	23.3	24.0	82.8	5.0	11.5	1380.0
	C9	7614	19 × 17	0.0	5.0	11.3	20.0	21.3	23.0	23.0	79.3	5.0	11.5	1322.5
Jute	J1	2961	10 × 16	1.0	7.0	18.0	24.3	28.0	28.0	28.7	98.8	5.0	12.2	1748.3
	J2	2961	10 × 12	1.0	7.0	18.0	24.0	27.3	29.0	29.0	100.0	5.6	11.7	1900.0
	J3	2961	10 × 8	0.0	6.3	17.7	23.0	26.0	27.0	29.0	100.0	6.0	11.2	1948.8
	J4	4773	10 × 16	2.0	10.7	20.3	24.0	27.0	27.0	28.0	96.6	5.6	11.1	1740.5
	J5	4773	10 × 12	1.0	11.0	17.0	22.0	25.3	26.0	27.3	94.2	6.3	10.1	1739.0
	J6	4773	10 × 8	1.0	9.0	16.7	22.0	25.0	27.0	27.0	93.1	6.6	9.8	1746.4

(a)

(b)

Fig. 8 Germination test in coir and jute RECSs: **a** high restriction for vegetation growth in coir RECSs; **b** less restriction for vegetation growth in jute RECSs and germination vegetation pass across the jute RECS yarns

softer jute fibers. In jute RECSs, it is observed that the germinating vegetation is able to pass across the jute yarns easily due to its finer and softer fibers, but in coir RECSs with rigid coir fibers, initial vegetation has to face higher restriction while germinating (Fig. 8).

At the end of 21st day, average number of primary roots is observed to be higher for the RECSs with higher weft density, and it is due to its better surface cover and moisture-holding capacity. Due to the same reason and due to additional advantage of lower flexural rigidity, jute RECSs show higher number of primary roots as compare to coir RECSs. At the end of 21st day, average root length is observed to be longer for the lower weft density RECSs; it is due to the non-availability of water on soil surface (lower moisture-holding capacity). Among different RECSs, it is observed that the coir RECSs has the longest root length at the end of 21st day. It is due to its higher flexural rigidity and lower moisture-holding capacity. Due to the poor availability of water on the soil surface, roots penetrate deeper into the soil. At increased weft density, total rooting of finer yarn coir RECSs and all jute RECs is higher due to higher percentage cover and better rooting, whereas in coarser yarn count RECSs total rooting reduces at increased weft density.

4 Conclusion

Influence of weft density on erosion control performance of coir and jute RECSs is studied at different slope angles and infiltration conditions along with its germination performance evaluation. From the study, it is observed that the increase in weft density results in decreased runoff volume and improved erosion control performance of jute RECSs at all slope angles (15°, 30° and 45°), at soil infiltration condition. It is due to the increased storage volume and higher restriction against the runoff water, and also due to the better drapability of jute RECSs. However, in coir RECSs, increase in weft density resulted in better performance at lower slope angles, but at higher slope angles erosion control percentage reduces in case of coarser weft yarn counts, it is due to its higher flexural rigidity. Similar to soil infiltration condition, jute RECSs with higher weft density has shown better performance at zero-infiltration condition, but the results of coir RECSs are inconsistent at slope angles due to the rigid protruding coir fibers on its surface. Weft density influences the germination performance of RECSs by varying the fractional cover and flexural rigidity of RECSs. Percentage of vegetation is observed to be higher at increased weft density in finer count RECSs, whereas at coarser count RECSs percentage of vegetation decreases at higher weft density, due to higher fractional cover and higher rigidity. At increased weft density, average number of roots germinated per plant are higher, but the average length of roots is lower. Therefore, at the end of 21st, total rooting of finer yarn count RECSs and jute RECSs is observed to be higher at increased weft density, whereas in coarser yarn count RECSs total rooting is observed to be lower at increased weft density. Among coir and jute RECSs, jute RECSs with soft and finer jute fibers facilitate better and faster growth of vegetation as compared to rigid coir RECSs.

References

1. Zuzel JF, Pikul JL (1993) Effects of straw mulch on runoff and erosion from small agricultural plots in northeastern Oregon. Soil Sci 156:111–117
2. Allen SR (1996) Evaluation and standardization of rolled erosion control products. Geotext Geomembr 14:207–221
3. Sutherland RA (1998a) A critical assessment of the research conducted at the hydraulics and erosion control laboratory—a focus on rolled erosion control systems applied to Hillslopes. Geotext Geomembr 16(2):87–118
4. Pimentel D (2006) Soil erosion; A food and environmental threat. J Environ Develop Sustain 8:117–119
5. Ogbobe O, Essien KS, Adebayo A (1998) A study of biodegradable geotextiles used for erosion control. Geosynthet Int 5(5):545–553
6. Hytiris N, McKay DJ, Addison PS (2001) A canopy flow equation for erosion control geosynthetics. Geosynthet Int 8(3):221–232
7. Sanyal T (2006) Jute Geotextiles in erosion control & strengthening of sub-grades-two case studies. Indian Chap Int Geosynth Soc CBIP 298:235–244
8. Sanyal T, Khastagir AK, Choudhry PK (2013) Design and development of improved varities of open weave jute geotextiles for erosion control. Indian J Geosynthet Ground Improve 2:9–13
9. Sutherland RA (1998b) Rolled erosion control systems for hillslope surface protection: a critical review, synthesis and analysis of available data. I. background and formative years. Land Degrad Dev 9:465–486
10. Sutherland RA (1998c) Rolled erosion control systems for hillslope surface protection: a critical review, synthesis and analysis of available data. II. The post-1990 period. Land Degrad Dev 9:487–511
11. Balan K (2004) Coir geotextiles- an eco-friendly material & Coir Geotextiles in watershed management and soil conservation. In Banerjee PK, Ramana GV, Shahu, JT, Rao GV (Eds), Geosynthetic new horizons. Asian books private limited, Papers of Geosynthetic India 2004 workshop held in New Delhi, India
12. Bhatia SK, Smith JL, Lake D, Walowsky D (2002) A technical and economic evaluation of geosynthetic rolled erosion control products in highway drainage channels. Geosynthet Int 9(2):125–148

13. Sutherland RA, Ziegler AD (2007) Effectiveness of coir-based rolled erosion control systems in reducing sediment transport from hillslopes. Appl Geogr 27:150–164
14. Gosh SK, Bhattacharyya R, Gupta KR (2015) Design and engineering of open weave jute soil saver for potential application in the field of soil erosion control and hill slope management. J Natural Fibers 12(6):561–573
15. Midha VK, Suresh KS (2013) Influence of woven structure on coir rolled erosion-control products. Geosynthet Int 20(6):396–407
16. Midha VK, Suresh KS, Sharma A (2014) Performance of transesterified jute rolled erosion control products. Geosynthet Int 21(5):301–309
17. Kalibová J, Jacka L, Petr˚u J (2016) The effectiveness of jute and coir blankets for erosion control in different field and laboratory conditions. J Solid Earth 7:469–479
18. State of environment—Atlas of India (State of Environment (SoE) Reports, 2011). https://www.soeatlas.org/Reports.aspx. Accessed 1 may 201
19. Fan JC, Wu MF (2001) Sustaining the global farm. Effect of soil strength, texture, slope steepness and rain fall intensity on interrill erosion of soils. In: Taiwan, Mohtar RH, Steinhardt GC, Stott DE (Eds), Scientific content-Selected papers from the 10th International Soil Conservation Organization Meeting, pp 588–593. Purdue University, Indiana, U.S (24–29 May 1999)
20. Shah DL, Shroff AV (2003) Soil mechanics and geotechnical engineering. AA Balkema Publisher, Tokyo, Japan, pp 120–150
21. Gardner CMK, Laryea KB, Unger PW (1999) Soil physical constraints to plant growth and crop production, pp 7–41. US Department of Agriculture, Bushland, Texas, USA
22. Cazzuffi D, Corneo A, Crippa E (2007) Slope stabilisation by perennial gramineae in Southern Italy: plant growth and temporal performance. Eco-Ground Bio-Eng 103:112–124
23. Blum A (2010) Plant Breeding for water- limited Environments. Springer Publishing, New York, USA, pp 12–44
24. Morgan RCP (2005) Soil erosion and conservation. Blackwell publishing ltd, Victoria, Australia, pp 11–42
25. Thomson JC, Ingold TS (1986) Use of Jute Fabrics in Erosion Control. In: Report to the Jute Market Promotion Project: International Jute Organization. Geneva: UNCTAD/GATT
26. Desai AN, Ravi K (2016) Geotextile made from natural fibre. MK Robert, Editor, Geotextiles, Woodhead publishing Ltd. Cambridge, UK 2005:61–86
27. Coir Development Board (2010) Coir Bhoovastra—some thoughts of marketing promotion. Coir Development Board, Allpey, Kerala, India, Technical Papers Collected from Authorized Sources 2010:1–30
28. John NWM (1996) Geotextiles. Blackie Academic and Professional, London, UK, pp 78–85
29. Mathi PM (2005) Coir. Frank, R.R., Editor, Bast and other plant fibres, Woodhead publishing Ltd. Cambridge, UK 2005:274–314
30. Deb MK, Chakrabarti S (2010) Construction of an economic road using jute geo-textiles in flexible pavements-an experimental study. 6th International Congress on Environmental Geotechnics, 8–12 November 2010. New Delhi, India, pp 879–886

Effect of Aerosol Charging on Energy Consumption During Pulse Jet Filtration Using Conductive Media

Arunangshu Mukhopadhyay, Sudev Dutta, A. K. Choudhary, and C. C. Reddy

1 Introduction

Minimizing energy consumption has developed into an important objective for various industrial sectors worldwide. To reduce power consumption and achieve better sustainability in filtration industry various systems have been designed globally [1–5]. One of the designs which is gaining popularity in recent times is pre-charger followed by pulse jet system [6]. Previous research analysed the filtration characteristics of conductive filter media on system assisted with pre-charger [4, 5, 7] but till date no study has been reported on energy consumption analysis using different types of conductive filter media as it can be a vital parameter for any industry using fabric filters. During their study four types of filter materials viz. PTFE coated media, Stainless Steel Fibre blended with PET media, Stainless Steel Scrim media and Carbon Filament Scrim Media were investigated at three levels of pre-charging viz. 4, 8, and 12 kV, and without charge. Among all the materials the emission and residual pressure drop results were found minimum for PTFE coated media. The cleaning of materials was done on pressure-based method. Although residual pressure drop indicates energy used by the system, but it is only a relative parameter and it does not include the other aspect of energy such as compressed energy due to pulsing and pre-charging, therefore considering all aspects while calculating energy it is also important to get the holistic view.

A. Mukhopadhyay (✉) · S. Dutta · A. K. Choudhary
Dr. B.R. Ambedkar, National Institute of Technology, Jalandhar 144011, India
e-mail: arunangshu@nitj.ac.in

C. C. Reddy
Indian Institute of Technology Ropar, Ropar 140001, India

V. Midha and A. Mukhopadhyay (eds.), *Recent Trends in Traditional and Technical Textiles*, https://doi.org/10.1007/978-981-15-9995-8_2

2 Materials and Method

The present study aims at analyzing the energy utilization behavior for five types of conductive filter media viz. PTFE coated media, Stainless steel fibre blended with PET media, Stainless steel scrim media, Carbon fibre blended with PET media and Carbon filament scrim media at three different levels of pre-charging viz 4, 8, and 12 kV, and without charging. The experiments were carried on laboratory-based pulse jet flat test rig set-up embedded with pre-charger (Fig. 1). The set-up comprised of a dust feeder for uniform dust feeding followed by a pre-charger installed to charge the aerosol particles. A dust layer forms on the surface of filter media during filtration, this dust layer is dislodged time to time on pressure-based method at peak pressure level of 1000 Pa through a pulsing time of 50 ms. The emission results were captured by an online particle size analyzer "Promo 2000" connected to downstream which works on the principle of light-scattering aerosol spectrometer system. The downstream side is attached to an online particle size analyzer "Promo 2000" is connected at the downstream side to analyze the emitted particles. The standard followed was ISO 11057 and the aerosol used was fly ash. Table 1 shows the material specifications.

Testing condition:

- Inlet face velocity: 2 m/min
- Air to cloth ratio: 2
- Inlet dust concentration: 50 g/m^3
- Tank pressure: 2 bar
- Valve opening time: 50 ms

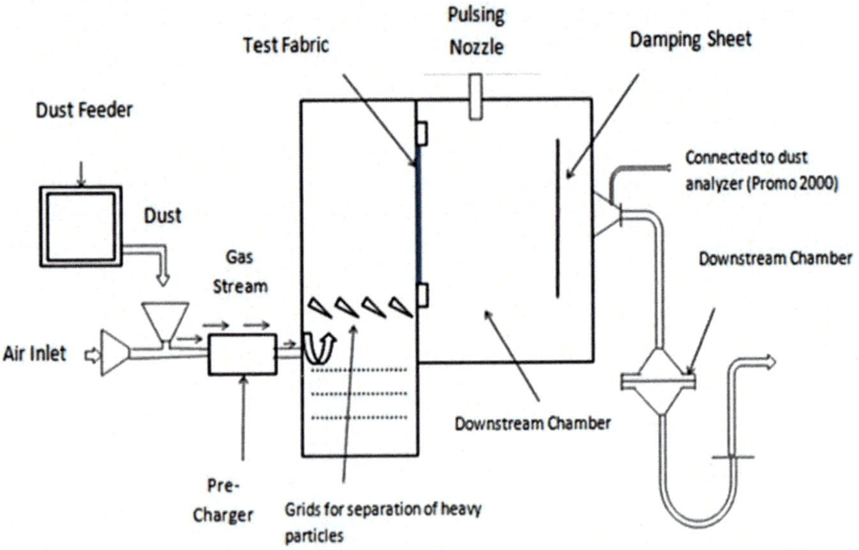

Fig. 1 Experimental set-up [4, 5, 7]

Table 1 Material specification[a]

Material type	Fabric mass (g/m^2)	Thickness (mm)	Air permeability ($m^3/m^2/min$)
PTFE coated filter media	550	2.3	18
S.S. fibre blended with PET media	550	2.3	15
S.S. scrim media	500	2.1	12
Carbon fibre blended with PET media	500	2.1	12
Carbon filament scrim media	500	2.1	12

[a]Material specifications except carbon fibre blended with PET media has been extracted from authors previous papers [4, 5, 7]

- Total filtration area: 0.09 m^2
- Pulsing at 1000 Pa differential pressure drop.

Testing sequence:

- Conditioning 30 cycles, cleaning pulse at 1000 Pa.
- Ageing 2500 cycles with a cleaning cycle at 20 s.
- Stabilizing 10 cycles, cleaning pulse at 1000 Pa
- Measuring for 2 h at 1000 Pa (Pressure based cleaning).

Specimen dimension:
50 cm × 18 cm.

Table 2 represents the emission and pressure drop values for final two hours of measuring phase of five types of conductive materials viz. PTFE coated media, Stainless steel fibre blended with PET, Stainless steel fibre scrim media, Carbon fibre blended with PET, and Carbon filament scrim media. The data has been extracted for final two hours of measuring phase. The filtration performance is found superior at higher charge level. Better performance at charged condition can be attributed to the effect of charge causing the aerosol particles to agglomerate and preventing depth filtration thus reducing the pressure drop and emission. Authors previous papers [4, 5, 7] reported that among all the materials filtration parameters in terms of residual pressure drop and downstream emission viz. $PM_{2.5}$, mass concentration and number concentration has been found to be the best for PTFE coated media followed by Stainless steel fibre blended with PET media, Stainless steel scrim media and Carbon filament scrim media. Charge dissipation test was carried out for all the materials. The media having lower charge dissipation exhibited better performance, this is because during charge dissipation the agglomerated particles break and loosen up, and more dissipation time suggests that agglomerated particles loosen slowly, which delays the particles to penetrate into the inner layer and settle on the surface of media, resulting in better surface deposition therefore, the emission reduces. Carbon filament scrim media exhibited highest charge dissipation behavior, therefore the residual pressure drop and emission results for this material have been the highest

Table 2 Material performance under different dust charge level[a]

Material type																				
Responses	PTFE coated filter media				Stainless steel fibre blended with PET media				Stainless steel fibre scrim media				Carbon fibre blended with PET media				Carbon filament scrim filter media			
	Dust charge level				Dust charge level				Dust charge level				Dust charge level				Dust charge level			
	0 kV	4 kV	8 kV	12 kV	0 kV	4 kV	8 kV	12 kV	0 kV	4 kV	8 kV	12 kV	0 kV	4 kV	8 kV	12 kv0 kV	0 kV	4 kV	8 kV	12 kV
$PM_{2.5}$ ($\mu g/m^3$)	811	690	605	514	919	734	620	576	1071	897	679	496	1181	944	748	592	1261	1089	893	702
Mass conc. (mg/m^3)	99	84	67	51	108	91	78	60	110	97	83	67	113	98	87	77	121	109	97	90
No. conc. (P/cm^3)	4006	3011	2716	2144	4567	3327	3016	2609	4975	4312	3179	2356	5178	4435	3926	3965	5699	5027	4178	3994
Residual pressure drop (Pa)	571	513	470	426	617	566	524	458	716	628	594	517	723	641	589	544	791	718	642	597

[a]Part of the date has been extracted from author's previous papers [4, 5, 7]

among all four materials, dissipation rate for stainless steel scrim media was lower compared to carbon filament scrim media and it exhibited lower emission and residual pressure drop values compared to stainless steel scrim media. PTFE coated material exhibited lowest dissipation rate as a result the filtration parameters were found to be best for it. In the reported study it has also been stated that reason for lower residual pressure drop for PTFE media is due to PTFE coating acting as a primary layer which facilitates surface deposition of particles allowing their easy dislodgement during pulsing. Apart from charge the intrinsic characteristics of material and its construction is also responsible for reduced energy, hence likewise charged condition for non charged condition also the behavior of all the materials is similar as that for charged condition.

The total energy consumption for final 2 h of measuring phase has been calculated which includes the power consumed by suction of gas, air compression required for media cleaning, and dust charging [8]. The fan energy or the gas suction can be calculated by using the measured pressure drop, total filtration time and gas flow rate (Eq. 1). The energy consumed for media cleaning was obtained by estimation of compressed air requirement (Eq. 2). Energy due to charge can be calculated by using the voltage and current applied to the pre-charger (Eq. 3).

$$\text{Energy of Fan(Kwh)} = \frac{q \times \mathrm{d}p \times t}{\eta \times 1000} \tag{1}$$

$$\text{Energy of Compressor (Kwh)} = \frac{N \times P \times Q\text{air}}{\eta \times 1000} \times 2.778 \times 10^{-7} \tag{2}$$

$$\text{Energy of Charge(Kwh)} = V \times \mathrm{I} \times t \times 2.778 \times 10^{-7} \tag{3}$$

where,

- q air flow rate (m^3/min.)
- η fan efficiency (%) considered as 85%
- N number of cleaning pulses
- P pulse pressure (Pa)
- Q_{air} discharge of air from reservoir at the time of pulsing (m^3)
- n exponent of polytrophic process considered as 1.33
- V voltage applied
- I current (Amperes)
- t operating time (hr.).

$$\text{Total Energy Consumption } E_{\text{Total}} \text{ in Kwh} = \text{Energy consumption of fan } (E_{\text{Fan}}) \text{ in Kwh} + \text{Energy consumption of compressor } (E_{\text{Comp.}}) \text{ in Kwh} + \text{Energy consumption of charge } \left(E_{\text{Charge}}\right) \text{ in Kwh} \tag{4}$$

Further to study the energy consumption behavior in realistic situation the energy consumption has been calculated considering a full scale bag house of 100 bags assuming 10 rows having 10 bags in each row. The bag dimensions were considered as 150 mm diameter and 6000 mm length. The values of air flow rate, discharge of air, number of pulses and current were scaled up for 100 bags using Eqs. 1, 2, and 3. For calculating fan energy the airflow rate (q) has been considered as 560 m^3/min for assuming same air to cloth ratio and residual pressure drop to that of flat media.

The compressed air requirement for each row is 10 times to that of single bag. It is assumed that same pulse intensity is used for each bag is same as that of flat specimen. In view of the above, the discharge of air (Q_{air}) has been considered 100 times to that of flat media set-up.

Energy required for charge has been calculated by proportionally scaling up the current (I) required per gram with rise in amount of inlet dust grams per minute. The current required for charging per gram of dust is 0.002 A per minute. For 100 bags the amount of inlet dust was 28,000 g/min. Therefore, total current (I) supplied for charging 28,000 g/min of dust has been 56 A.

3 Results and Discussion

It has been observed that energy consumption for the fan has been found to be the highest among fan, compressor and charge for all materials. Higher contribution of fan in overall energy consumption is because fan operates throughout the filtration process as it serves the purpose of gas suction and directs the gas flow over the surface of the filter media. As more layers of dust get deposited over the media surface higher fan energy is required for gas suction, while compressor energy is required only at the time of pulsing therefore its contribution is comparatively less in overall power consumption. Figure 2 shows the trend of energy consumption by fan for all five materials at various dust charge levels. For dust charging the overall drop in energy consumption of fan is noted to be around 10–14% compared to uncharged condition. It can be observed from the graph that there is a sharp decrease in slope as the pre-charging level rises from 0 to 4 kV, further from 4 to 8 kV and 8 to 12 kV a stability in trend is noted for all materials.

It can also be noted that as the dust charge level rises from 0 kV to 4 kV the rate of decrease in fan energy is higher as compared to rise in charge from 4 kV to 8 kV and 8 kV to 12 kV respectively. This is because at uncharged condition there are more number of smaller particles in the flue gas approaching the media which affects the surface deposition over the media surface and results in nonuniform cake formation as due to number of smaller particles there are chances that some of the particles may penetrate inside the media, therefore high fan energy is required to maintain a nominal flow throughout the surface of the media, while at 4 kV there are less number of smaller particles in the gas steam due to the effect of agglomeration the particle size get larger which improves the deposition of dust over the media surface

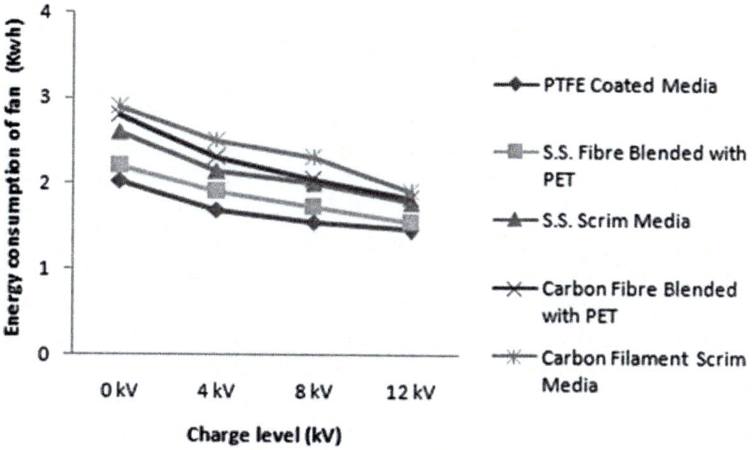

Fig. 2 Energy consumption trend of fan for all materials

resulting in a uniform cake layer formation due to this an easy nominal gas flow can be maintained throughout the media surface at relatively lower fan energy.

Effect of agglomeration will not be to the similar extent as charge increase from 4 to 8 kV and further. This is because at 8 kV and further the particle diameter of the agglomerated particles is likely to be larger compared to 4 kV which is expected to contribute in formation of a little better cake due to larger particles. Therefore for 8 kV and further energy required by the fan for maintaining a throughout optimum gas flow will be less compared to 4 kV but it will not be that much less as it got reduced from without charge to 4 kV, this is because for each charge level uniform surface deposition is already accomplished, for 8 kV and further the cake deposition is expected to be little better which can contribute in relatively reduced fan energy, but the rate of reduction will not be that much higher. This also gives an indication that effect of charge impactful only up to a certain limit.

Another notable point has been decrease in difference in emission between materials at higher pre-charging level compared to uncharged condition which indicates that pre-charging is more effective for materials that exhibit higher emission at uncharged condition, as higher emission levels will likely to have more number of smaller particles, the effect of agglomeration is more impactful on smaller particles [6], therefore at higher charge a large improvement can be noticed. As already stated in materials and method section the filtration parameters in terms of pressure drop and downstream emission viz. $PM_{2.5}$, mass concentration and number concentration has been found to be best for PTFE coated media followed by stainless steel fibre blended with PET media, stainless steel scrim media, and carbon filament scrim media in previously reported studies. The rank wise performance of materials in terms energy consumption is similar to the material performance in terms of emission and residual pressure drop for the reported studies [4, 5, 7]. Among all the materials the energy consumed is found to be least for PTFE coated filter media followed by stainless steel fibre blended with PET media, stainless steel scrim media, carbon fibre blended with PET media and lastly carbon filament scrim media which exhibited the highest energy consumption for fan.

Figure 3 depicts the trend for energy utilized by compressor at various pre-charging levels for all the materials. Compressor energy indicates energy utilized during cleaning. Overall reduction in power utilization for compressor has been found to be ranging between 11 and 17% for charged conditions. From 0 to 4 kV the rate of reduction in compressor energy is higher compared to 4 to 8 kV and further, this is because as there are more number of fine particles in the gas stream at uncharged condition, which prevents smooth deposition of particles over the media surface because large number of particles are likely to penetrate within the pores of media, due to continuous penetration there comes a stage where all the pores get blocked resulting in early clogging of media which causes in more frequent pulsing resulting in higher compressor energy. At 4 kV the effect of agglomeration causes a bridging effect over the media surface as a result very few particles penetrate inside, hence the pulsing interval is increased which enables less compressor energy to be utilized. Similarly for higher charge level also the reason for reduced compressor energy is the effect of agglomeration.

The performance of materials has been in same order as that for fan energy utilization. The material wise performance for energy consumption of compressor has been similar as observed in fan energy. The reason could be attributed to charge dissipation test results and filtration parameter results as mentioned in fan energy section.

Figure 4 depicts the overall power consumption trend for all filter media. In case of charging the overall energy consumption is found to be reduced ranging from 5 to 9% compared to uncharged condition. The graph shows that a sharp decrease in slope from 0 to 4 kV and further a relatively stable trend can be noted from 4 to 12 kV which is much similar to the trends obtained for fan and compressor energy. The material wise ranking for overall energy consumption has been PTFE coated

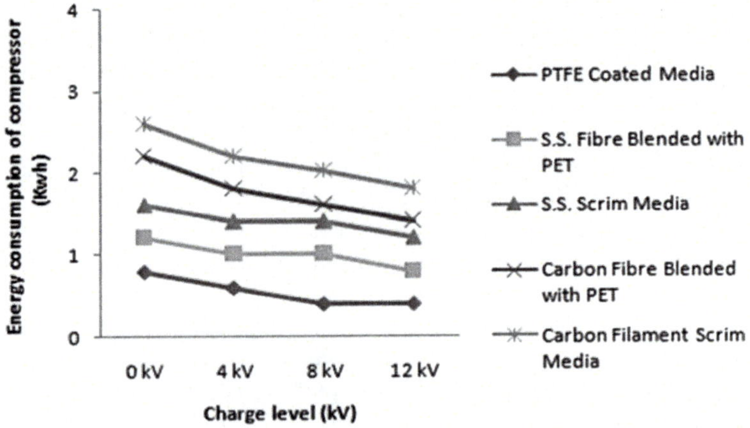

Fig. 3 Energy consumption trend of compressor for all materials

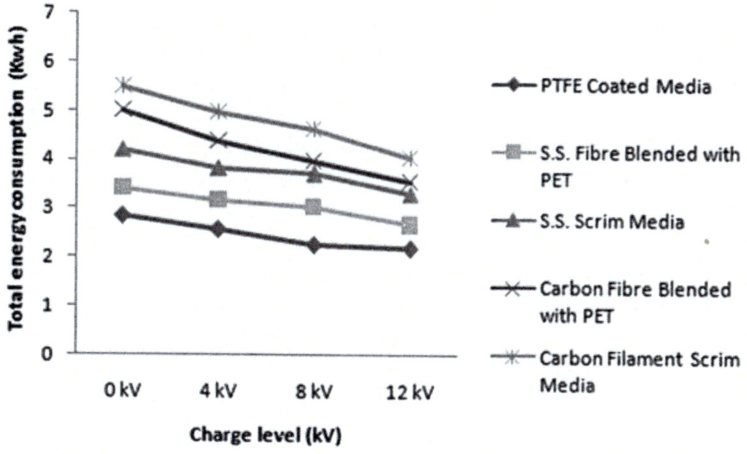

Fig. 4 Overall energy consumption trend for all materials

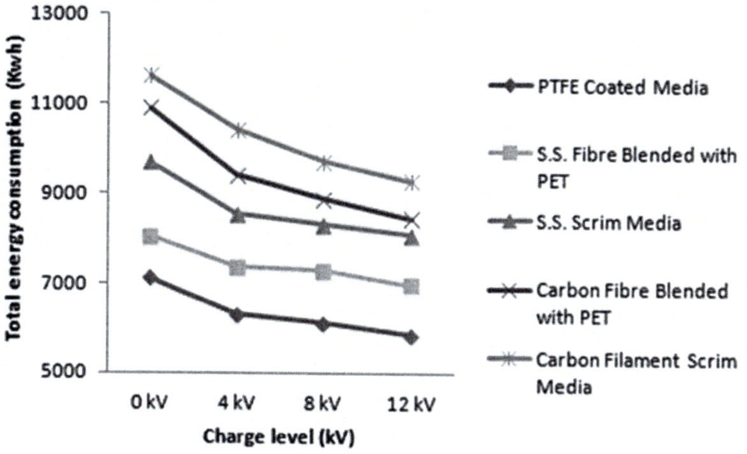

Fig. 5 Total energy consumption for 100 bags

filter media followed by stainless steel fibre blended with PET media, stainless steel scrim media, carbon fibre blended with PET media and lastly carbon filament scrim media which exhibited the highest energy consumption for fan. The reason has been similar to as stated for fan and compressor energy in case of charged as well as non charged condition.

To achieve better understanding and compare the energy consumption values with a full scale bag house condition the energy consumption is further calculated assuming 100 bags by considering the same pressure drop values as that in Table 2 [4, 5, 7] for calculation.

Figure 5 depicts the total energy consumption behavior for 100 bags. It can be noted that rate of reduction in power consumption is much higher for 0–4 kV as compared to 4–8 kV and further, which suggests that pre-charging of dust can be more effective in dealing with higher volume of gas. As already mentioned the effect of agglomeration is more impactful on fine particles [6], and more number of finer particles are expected to be present in flue gas stream for a full scale bag house. The rank wise performance has been similar to that for the tested results, i.e., PTFE coated filter media followed by stainless steel fibre blended with PET media, stainless steel scrim media, carbon fibre blended with PET media, and lastly carbon filament scrim media which exhibited the highest energy consumption for fan. The reason can be attributed to the charge dissipation behavior of materials which has already been discussed previously. The media taking longer time to dissipate charge is having better filtration results, the reason for this has already been discussed.

The reduction in power consumption at pre-charging level is due to cumulative behavior of charge and material. It has been observed that each material is behaving differently, this difference in behavior is due to intrinsic behavior of material as their structural properties are different and also due to charge. As far as only charge is concerned two best materials viz. PTFE coated media and stainless steel fibre blended with PET have been identified from analysis, therefore ANOVA has been carried out for full scale bag house considering these two materials (Table 3). The calculated F value is higher than the table value for each case which indicates that each factor and their interaction is having a significant impact on energy consumption results.

Figure 6 depicts the proportionate contribution in total energy consumption at 0 and 4 kV charge for PTFE coated media. From the graph it can be seen that a drastic reduction in contribution of compressor for 100 bags in comparison to media tested in system assisted with pre-charger. Compressor energy is noted to reduce from 33 to 15% in case of 100 Bags, whereas contribution of fan energy elevated from 57 to 80% for 100 Bags, this can be related to the fact that larger filtration area requires higher suction energy for maintaining nominal flow of gas throughout the filter media as compared to smaller filter area. For compressor energy, the difference in percentage contribution is comparatively less and for energy due to charge, there is not much variation in percentage contribution for 100 bags in comparison with flat media test rig.

Table 3 ANOVA for energy consumption of 100 bags between PTFE coated and PET blended with S.S. fibre media

Sources	S.S.	d.f.	Mean square	$F_{cal.}$	$F_{tab.}$ at $\alpha = 0.05$	Contribution (%)
Material	108713685	1	108713685	185.5	5.31	58
Charge	53419655	3	17806552	30.4	4.06	28.5
Interaction	20618113	3	6872704	11.8	4.06	11
Residual	4685935	8	585742			2.5
Total	187437388	15				

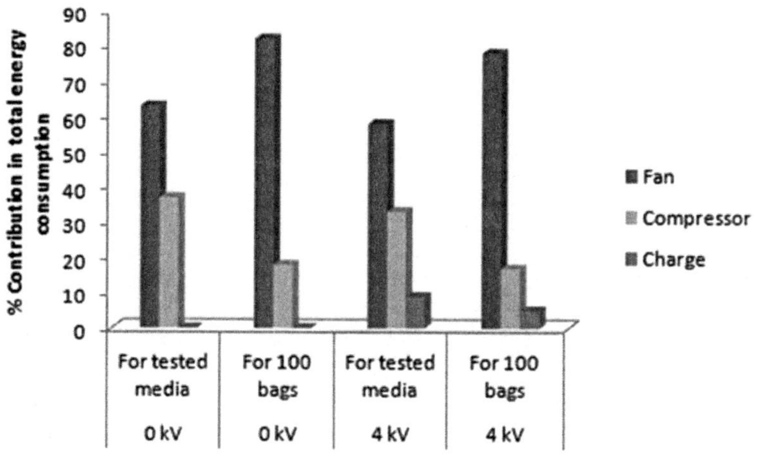

Fig. 6 Proportionate contribution in total energy consumption at 0 and 4 kV charge for PTFE coated material

4 Conclusions

1. Overall energy consumption is reduced with the increase in charge level in the pre-charger till 12 kV. Cumulative contribution of material type and charging has been observed in reducing power consumption at pre-charging levels.
2. The decrease in power consumption is higher at initial levels of pre-charging from 0 to 4 kV which suggests that effect of charge is more effective at initial level.
3. Surface properties of media is an important factor as far as energy aspect is concerned, from the charge dissipation behavior of materials, it is clear that media taking longer time to dissipate charge is exhibiting less energy.
4. The percentage reduction in power consumption from 0 to 4 kV is much higher for full scale bag house which indicates that pre-charging is more beneficial for smaller particles, as full scale bag house will have relatively more number of smaller particles.
5. There has been a drastic increase in percentage contribution of fan energy for full scale bag house, While for compressor energy, the difference in percentage contribution is comparatively less and for energy due to charge, there is not much variation in percentage contribution.

References

1. White HJ (1974) Resistivity problems in electrostatic precipitation. Air Pollut Control Associat 24(4):314–338
2. Masuda S, Washizu M (1979) Ionic charging of a very high resistivity spherical particle. J Electrostat 6(1):57–67
3. Lin WY, Chang YY, Lien CT, Kuo CW (2011) Separation characteristics of submicron particles in an electrostatic precipitator with alternating electric field corona charger. Aerosol Sci Technol 45:393–400
4. Dutta S, Mukhopadhyay A, Choudhary AK, Reddy CC (2018) Hybrid aerosol filtration systems—a review. Int J Eng Technol Sci Res 5(5):535–555
5. Dutta S, Mukhopadhyay A, Choudhary AK, Reddy CC (2018) Response of conductive and PTFE-coated polyester filters to cold fly ash at different dust charge levels. J Institut Eng (India) Series E99(2):219–226. https://doi.org/10.1007/s40034-018-0129-0
6. Krupa A, Jaworek A, Marchewicz A, Sobczyk A, Czech T, Antes T, Śliwiński Ł, Ottawa A, Szudyga M, Charchalis A (2017) Submicron particles emission control by electrostatic agglomeration. przegląd elektrotechniczny 93:237–241
7. Dutta S, Mukhopadhyay A, Choudhary AK, Reddy CC (2019) Filtration behaviour of polyester conductive filter mediaon pulse jet test rig assisted with pre-charger. J Institut Eng (India): Series E100(1):47–57. https://doi.org/10.1007/s40034-019-00139-z
8. Choi HK, Park SJ, Lim JH, Kim SD, Park HS, Park YO (2002) A study on the characteristics improvement in filtration performance by dust pre-charging. Korean J Chem Eng 19(2):342–346

Thermal Contact Properties of Rib Cotton Weaves at Increased Moisture Level

Lubos Hes

1 Introduction

Thermal properties of textiles and such as thermal resistance, thermal conductivity and thermal absorbtivity present principal thermal comfort parameters of textile fabrics and clothing and their determination and applications in functional clothing is paid high attention in recent decades. Whereas thermal conductivity and thermal resistance are well known parameters as well as the related measuring methods and instruments, thermal absorbtivity b [$Ws^{1/2}/m^2K$] of fabrics was introduced in textiles quite recently, in 1987 by Hes [1]. This parameter characterising thermal feeling (heat flow level) during short contact of human skin with a fabric surface. For short time of thermal contact τ between the human skin and the fabric, the measured fabric can be simplified into semi-infinite homogenous body with thermal capacity ρc [J/m^3] and initial temperature t_2. Unsteady temperature field between the human skin (with temperature t_1) and fabric with respect to of boundary conditions offers a relationship, which enables to determine the heat flow q [W/m^2] course passing through the fabric:

$$q = b(t_1 - t_2)/(\pi\tau)^{1/2}, \quad b = (\lambda\rho c)^{1/2}\left[Ws^{1/2}/m^2K\right]$$

where ρc [J/m^3] its thermal capacity of the fabric and the term b presents **thermal absorbtivity** of fabrics. The higher is thermal absorbtivity of the fabric, the cooler is its feeling. In the textile praxis this parameter ranges from 20 [$Ws^{1/2}/m^2K$] for fine nonwoven webs to 200 [$Ws^{1/2}/m^2K$] for standard cotton weaves up to 600

L. Hes (✉)
Department of Textile Evaluation, Faculty of Textile Engineering, Technical University of Liberec, 46117 Liberec, Czech Republic
e-mail: lubos.hes@gmail.com

V. Midha and A. Mukhopadhyay (eds.), *Recent Trends in Traditional and Technical Textiles*, https://doi.org/10.1007/978-981-15-9995-8_3

$[Ws^{1/2}/m^2K]$ for wet fabrics. Determination of this parameter of fabrics enables the ALAMBETA computer-controlled non-destructive tester [2].

In the study [3], this instrument was used to determine the above explained thermal insulation and thermal contact properties of 12 cotton rib fabrics, with the objective, to determine the effect of the reduced thermal contact area of these fabrics on their thermal properties. Moreover, for the first time also the effect of the increased moisture on the mentioned thermal parameters of these fabrics was investigated, as papers on thermal properties of wet fabrics are quite rare [4]. For comparison, also thermal properties of smooth cotton velvet fabrics with similar square mass were measured. In this paper, only the effect of the added moisture and relative contact area on thermal contact properties of the mentioned rib fabrics will be presented.

2 Experimental

See Fig. 1 and Table 1.

The samples were washed, dried, weighed and then tested at the laboratory conditions. Then, the moisture was added and the samples were stepwise dried.

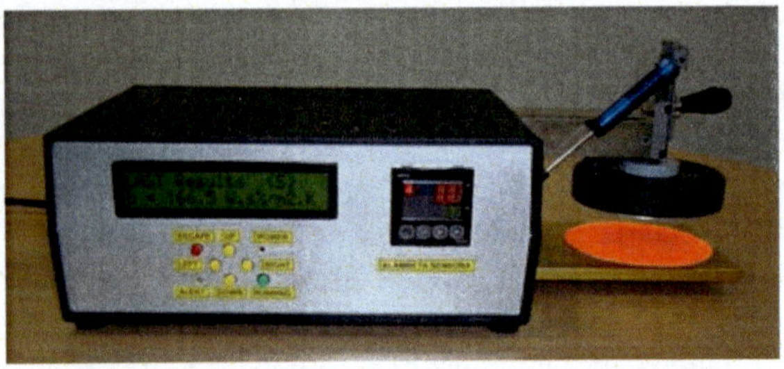

Fig. 1 ALAMBETA thermal tester. Besides thermal absorbtivity this apparatus used in this study enables also the measurement of thermal conductivity, thermal resistance and sample thickness

Table 1 Woven samples tested by the ALAMBETA at the contact pressure 200 Pa

Sample No.	Composition	Sq. mass m_s [g.m^{-2}]	Relative rib width H_r [-]	Sample No.	Composition	Sq. mass m_s [g.m^{-2}]	Relative rib width H_r [-]
1 A	Corduroy	219	0,6	8	Corduroy	304	0,8
2 E	Corduroy	320	0,6 E	9	Corduroy	316	0,8
3 A	Corduroy	178	0,7	10	Corduroy	336	0,8
4	Corduroy	280	0,7	11	Corduroy	327	0,9
5 E	Corduroy	313	0,7	12	Corduroy	329	0,9
6	Corduroy	338	0,7	S1 A	Velvet	284	1
7	Corduroy	364	0,7	S2	Velvet	333	1

3 Results and Discussion

See Figs. 2 and 3.

From the presented results follows, that thermal absorptivity of the studied fabrics increases with the moisture level. However, relative heavy rib fabrics with low thermal

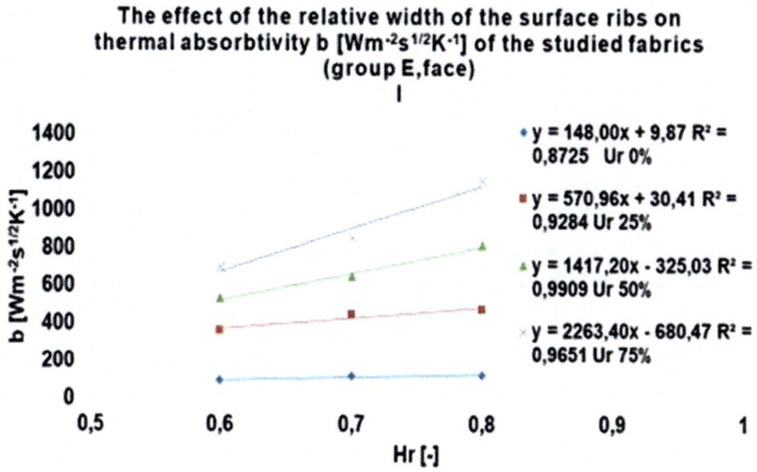

Fig. 2 As presented, thermal absorptivity of the samples, group A, increases linearly with the level of the added moisture. Higher relative width of the rib causes higher values of thermal absorptivity

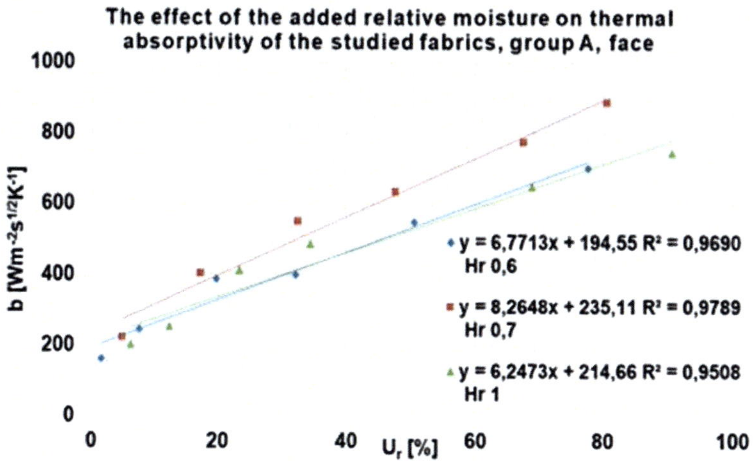

Fig. 3 As presented, thermal absorptivity of the samples, group A, increases linearly with the level of the added moisture. Higher relative width of the rib causes higher values of thermal absorptivity

contact area, may still offer dry contact feeling at the added moisture level 20–25%, which may correspond to the sweating caused moisture level at medium physical effort, within not too long time. Unfortunately, detailed studies on subjective evaluation of thermal contact feeling of underwear and shirt fabrics are still missing.

From the results presented in the study [3] follows, that also thermal conductivity of rib cotton fabrics increases with the moisture level, and decreases with the reduced thermal contact area. The complex results will be published later.

4 Conclusions

In the study, the effect of thermal contact area and moisture on thermal contact feeling of corduroy cotton rib fabrics was investigated. It was found, that thermal absorptivity of the studied fabrics increases with the moisture level and decreases with the relative contact area. Thus, at low/medium moisture levels, the corduroy fabrics, along with sufficient water vapour permeability (as follows from other studies), provide good thermal comfort, if the physical effort is moderate. That is why the corduroy fabrics are still often used in the work and casual clothing.

References

1. Bednarova M (2019) Thermal comfort properties of rib fabrics (in Czech), BSc Thesis, Technical University of Liberec
2. Hes L (1987) Thermal properties of nonwovens. In: Massenaux B (ed), Proceedings of the INDEX 1987 Congress, Section B1, Geneva
3. Hes L, Dolezal I (1989) New method and equipment for measuring thermal properties of textiles. J Text Mach Soc Jpn 42:T124–128
4. Schneider AM et al (1987) Heat transfer through moist fabrics. Ph.D. Thesis, UNSW Australia

Leveraging Artificial Intelligence to Foster Innovation and Inclusive Growth in the Textile Value Chain

Satyadev Rosunee

1 Introduction

Artificial Intelligence (AI) is powered by data. Large datasets improve and sharpen the output capabilities of AI systems. AI used intelligently can help the textile industry (both traditional and technical textiles) mine a lot of data to improve productivity, enable myriad forms of value addition, avoid machine down time, improve quality control, optimize processes and resource utilisation, smarten predictive maintenance, among other potential benefits. The above-mentioned areas are the 'low-hanging fruits' of research & development and innovation in the textile value chain.

One of the major weaknesses of textile plants is that they possess a lot of process and operations data, but have not yet invested in the tools to extract tangible business value. In textile production, AI will potentially offer enhanced flexibility, reactivity, forward planning and decision-making freedom on the shop floor as some decision-making responsibilities are transferred to the AI-enabled System. In short, AI has the capability to transform textile factories, making them more resource-efficient while also improving their environmental footprint.

This paper will also try to dispel the myth and misconceptions that AI will soon surpass human intelligence, somehow 'take over' the world and render billions of workers jobless. AI is not even close to replacing human ingenuity, empathy, creativity and imagination. AI will effectively replace jobs that follow a fixed pattern, are repetitive and are routine. It is just a matter of time before algorithms replace such jobs with phenomenal increase in productivity. Nevertheless, entirely new jobs would be created although some degree of re-skilling might be necessary to ensure best fit and inclusiveness. After all, the textile sector has since quite some time adopted state-of-the-art fully automated spinning plants, where human intervention has been

S. Rosunee (✉)
Faculty of Engineering, University of Mauritius, Reduit, Mauritius
e-mail: s.rosunee@uom.ac.mu

V. Midha and A. Mukhopadhyay (eds.), *Recent Trends in Traditional and Technical Textiles*, https://doi.org/10.1007/978-981-15-9995-8_4

drastically reduced. Both India and Mauritius have recently launched their national AI strategies. The strategy papers indeed offer good potential for collaborative research, development and innovation in the textile sector.

2 Methodology

The views and insights reported in this paper are based on semi-structured interviews and telephonic conversations carried out with the managing directors of six textile factories in Mauritius. In most cases, their views reflect their future investment plans in AI-driven technologies, mainly to improve profitability and environmental footprint.

3 Impacts of AI

The textile industry's product range is one of the most diverse in the manufacturing sector. The industry processes a variety of textile fibres; processing parameters and methods vary; the product mix is relatively high; value-added depends on end-use. With so many variables in the production set-up, the impact of AI is proving to be positive, wide-ranging and constructive. AI has nurtured a number of fast growing technology-enabled branches, Fig. 1. These established tools— machine learning (ML), machine vision, expert systems and robotics, etc.— will potentially augment

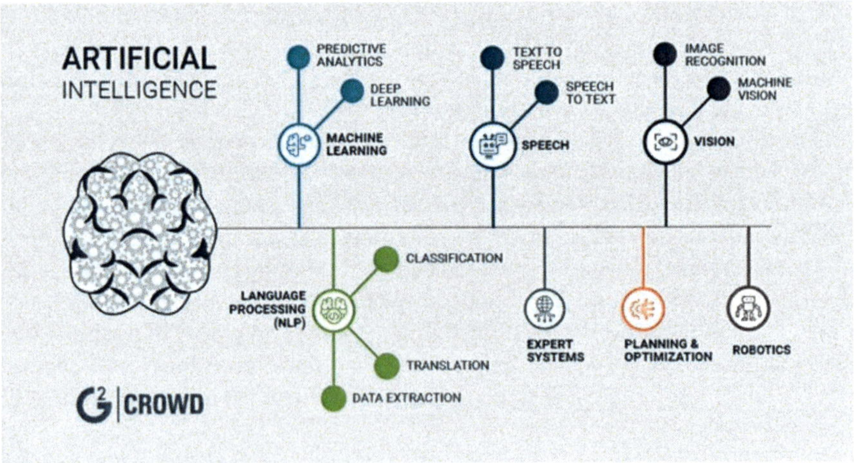

Fig. 1 Branches of AI (*Source* G2 Crowd)

the expertise of textile professionals and help them in their main mission, i.e. to produce according to specifications and fulfil customer requirements.

Myriad types of production/processes-related data that are deemed useless today and applications that are thought to be impossible are being enabled by sophisticated AI and ML algorithms. Smart algorithms are already delivering new insights about selection of raw materials, fabric construction parameters in order to engineer the thermo-physiological comfort properties of woven fabrics [1].

Thus the textile industry is expected to be a major beneficiary of AI based solutions, through flexible and adaptable technical systems to automate processes and machinery to respond to unfamiliar or unexpected situations by making processing decisions smarter. Other major potential impact areas include research & development, supply chain management (demand forecasting), and production as AI is poised to achieve cost reduction and increase productivity & efficiency.

According to market intelligence agency TrendForce (2019) [2], smart manufacturing supported by industrial IoT and AI is projected to grow noticeably in the next 3–5 years. The cost of smart manufacturing systems is expected to drop as the market grows from $200 billion in 2018 to over $320 billion by 2020. Another low-hanging fruit where AI can have major cost-benefit will be Plant Maintenance (predictive maintenance and reduced machine downtime) and increased asset utilization. Root Cause Analysis based on machine learning is a potential game changer as well. Given the variety of inputs in textile production (fibres, yarns, process parameters, water, dyeing and finishing chemicals, machine settings, etc.). ML-driven root cause analysis is a major area of interest for the Mauritian textile industry [3]. The directors interviewed reported that they are eager to collaborate with academia (local & overseas) in order to develop such a tool.

By 2030, all regions and sectors of the global economy will benefit from the tangible multiplier effects of AI. AI is expected to generate 15trillion USD of 'value-added' to the world economy, an amount greater than the combined GDP of China and India. By 2030, AI's value-added to the Chinese economy is expected to reach 7.0 trillion USD, while that of the USA will be about 3.7 trillion USD. Europe and developed Asia would also experience significant economic gains from AI [4]. Thus, the overarching impact of AI in textiles research and development is expected to be highly dynamic. AI will create entirely new areas of research that can help the textile industry, especially those operating in the developing world, to comfortably align with Sustainable Development Goals (SDGs). This paper will not over-emphasise on the socio-economic and environmental challenges posed by the textile industry in developing countries.

3.1 Real Time Fabric Inspection

Since AI's first successful commercial application was linked to pattern and image recognition, it is being adopted by Mauritian textile manufacturers for real-time fabric inspection. A simple system (adapted from existing commercial systems) is

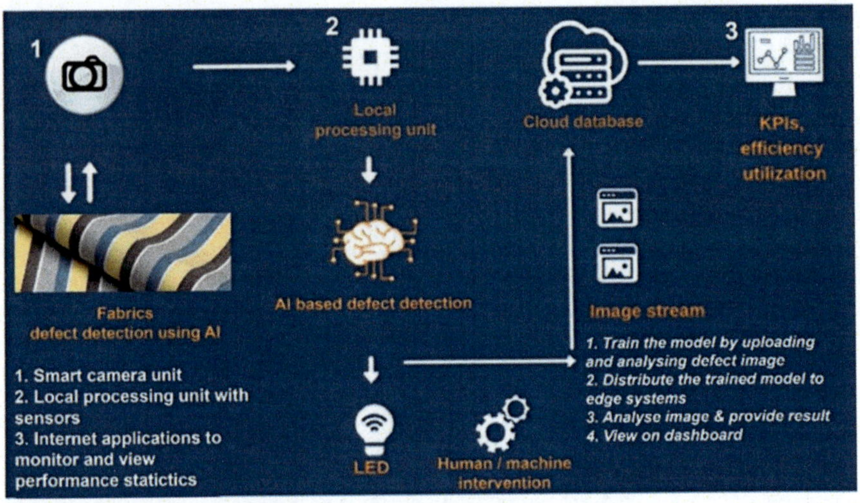

Fig. 2 Real-time fabric inspection

shown, Fig. 2. A smart camera unit acquires images of the fabric in real time. The images are processed by a local processing unit and then fed to the AI-driven fabric defect detection system. The system is equipped with a machine-learning algorithm, trained with a fairly comprehensive fabric data set, to identify defects and deviations from product specifications.

Real-time fabric inspection replaces operators' visual fabric inspection, which is error-prone mainly due to fatigue. Mauritian fabric manufacturers argue that it has the potential to reduce claims for compensation from garment manufacturers to whom they supply finished goods. Compensation claims are a perpetual drain on their profit margins, which is quite slim owing to global competition. It is to be noted that claims made by garment manufacturers can sometimes amount to 50% of the invoice value, which makes the acquisition of such a system cost-effective and pertinent to the survival of the business.

3.2 *Machine Maintenance*

A systematic machine maintenance schedule or plan is a key pillar of an efficient textile manufacturing plant. Mauritian manufacturers believe that adoption of machine learning and data analytics may result in 10% reduction in annual maintenance costs, up to 20% machine downtime reduction and 25% reduction in inspection costs through predictive maintenance.

3.3 Supply Chain Management

Machine learning (ML) is a tool that is being increasingly used for supply chain forecasting. ML is expected to reduce forecasting errors by 50%; consequently reducing costs related to transport and warehousing and supply chain administration by 5–10% and 25–40%, respectively. AI and machine-learning algorithms are being integrated into purchasing, strategic sourcing and cost management.

3.4 The Human Factor

Textile and garment manufacturing has historically been labour-intensive. The industry generally recruited among the low-skilled and provided training on-the-job. Over the years, low-skilled labour has helped to keep production costs down. The automation of production processes was gradual but has been accelerating in the last quarter of the twentieth century. However, recent advances in artificial intelligence and cheaper cost of computing power have made expert systems more powerful and more affordable. Still, a human is required to decide where and how AI is deployed, and which key decisions are turned over to AI. Although automation has made the process of turning data into useful information faster than ever, it has not done away with the human task of endowing information with meaning, and acting on that information creatively. Nevertheless, AI is poised to replace jobs that are routine, repetitive and follow a specific pattern. Textile factories would gain by kick-starting training and re-skilling programmes to boost employee motivation and productivity. The textile and garment industry in Mauritius has already embarked on such training and re-skilling programmes. Stakeholders have emphasised that prior to the adoption of AI technologies, it is crucial to develop a coherent communication strategy with employees so that AI is not perceived as a job security risk factor.

3.5 Conclusions

Artificial intelligence is a potential game changer. It might revolutionise the way we manufacture textiles at scale, improve productivity and reduce negative environmental impacts. It will inevitably trigger the creation of new business models, possibly the Ubers and Airbnbs of the textile business.

Employees will need to become 'AI literate', adapt to the new work environment and to anticipate how AI can transform their career and role in the company. In general people are getting accustomed to the idea that automated decision-making by algorithms is increasingly going to be part of our daily lives and work-life. Workers will need to have a working knowledge of how algorithms work and how automated decisions are made, so that they can flag biased decisions, which may arise if the AI system has been trained with compromised data.

Finally, given the multi-pronged nature of AI and the fact that it is driving so many other technologies, developing a robust ethical framework is an absolute necessity, mainly for privacy and data security. This framework must also deal with AI's impact on workers' rights, which is at the core of sound industrial relations in Mauritius.

References

1. Abel J (Oct 2019) Predictive modelling of thermal contact feeling & thermal physiological comfort parameters of single layer cotton woven fabric. Ph.D. Thesis, University of Mauritius
2. https://www.trendforce.com/research/artificial_intelligence
3. http://mtci.govmu.org/English/Documents/2018/Launching%20Digital%20Transformation%20Strategy%20191218/Mauritius%20AI%20Strategy%20(7).pdf (Mauritius Artificial Intelligence Strategy, 2018)
4. https://www.pwc.com/us/en/press-releases/2017/report-on-global-impact-and-adoption-of-ai.html (PWC Report, 2017)

The Effect of Structural Parameter on Pressure Behaviour of Tubular Bandage

Monica Sikka, Mamta Devi, and Samridhi Garg

1 Introduction

Compression therapy is the most widely used treatment for venous leg ulcer (VLU), and it has been utilised in different forms for more than four centuries [1]. The compression therapy can be given by different through compression bandages, compression hosiery; IPC (Intermittent pneumatic compression therapy), but out of these compression bandaging is widely used [2]. There are wide varieties of compression bandages available, according to the requirement of pressure, a suitable bandage has to be selected for the specific problem based on required pressure level. [3].

Different types of compression materials have different compression characteristic and the ability of a bandage to maintain pressure is determined by the elastomeric properties of the yarns and the structure of the fabric [4].

The main function of compression materials is to exert an external pressure on limb, which leads to better venous blood flow [5]. It is important to maintain the required level of pressure because excessive pressure can lead to tissue damage, pressure sores and necrosis. Reverse gradient compression is likely to worsen the condition as it increases the pressure in the veins. Limb damage or treatment failure may result in limb amputation [6]. A direct relation exists between extension and applied force. For the given compression materials, the higher the stretch percentage, the higher the pressure. The capacity and efficiency of a material to sustain pressure is greatly dependent on its ability to maintain this internal stress which depends on compression materials structure, material type, amount of elastane, etc.

M. Sikka (✉) · M. Devi · S. Garg
Department of Textile Technology, National Institute of Technology, Jalandhar 144011, India
e-mail: sikkam@nitj.ac.in

S. Garg
e-mail: samridhi700@gmail.com

V. Midha and A. Mukhopadhyay (eds.), *Recent Trends in Traditional and Technical Textiles*, https://doi.org/10.1007/978-981-15-9995-8_5

It is desired that the compression material should be able to preserve its compressive pressure even after being worn for an entire day and do not lose their elastic stretch recovery. In this respect knitted structure of bandage is more preferred. The flat and circular knitting is used as the knitting techniques to produce compression hosieries [7]. In circular knitting technology, the diameter of the machine and the number of needles is fixed during production of a particular product. These garments are mostly knitted as plain knitted structure. Although knitted compression material is good in sustaining the pressure for longer period but it shows viscoelastic properties and their recovery depends on the drawn ratio [5, 8]. The exerted loads in the production process and during application of fabrics also cause to stretch and let deformations in the fabric. In manufacturing of knitted bandage core spun cotton yarns, with elastane component in the core and cotton in the sheath, have become quite popular, because the cotton fibres covering the elastane core and provide the necessary tactile aesthetics to the wearer along with thermo-physiological comfort and performance characteristics as compared to the conventional yarns [9, 10].

In present work tubular knitted bandage is prepared using core spun yarn of varying lycra denier and its pressure behaviour was analysed at different stretch %, different limb circumference of different hardness. The interactive effect of these parameters on pressure behaviour is also analysed statistically.

2 Material and Methods

Cotton/lycra core spun yarn with three different lycra Denier, i.e 20, 30, 40 Denier. The physical properties of the tubular bandage are shown in Table 1.

The individual and interactive effect of lycra(denier), stitch length(mm), stretch (%), hardness of limb(degree) on tubular bandage pressure was studied at three levels

Table 1 Physical properties of the tubular bandage

	Lycra denier	Stitch length (mm)	Course and Wale per inch		Thickness (mm)	GSM of fabric
			Course	Wale		
Cotton/lycra core spun yarn (40s)	20	2.5	84	41	1.70	171
		3	76	37	2.23	169
		3.5	60	29	2.35	152
	30	2.5	80	40	1.68	205
		3	70	31	2.23	202
		3.5	56	26	2.58	188
	40	2.5	72	31	1.66	142
		3	56	26	1.51	135
		3.5	50	24	2.03	124

Table 2 Design plan of experiment

Factors	Levels
Lycra (Denier)	20
	30
	40
Stitch length (mm)	2.5
	3
	3.5
Stretch (%)	80
	100
	120
Hardness of limb (Degree)	10
	20
	30

using Full Factorial Design (34 = 81 Runs). Various factors have been considered for this study at yarn and fabric stage have been presented in Table 2.

In order to study the individual and interactive effect of each parameter on tubular bandage pressure, response surface regression equations were developed for tubular bandage pressure. The general relationship between the response Y (tubular bandage pressure) and the different parameters is given below as:

$$\begin{aligned} Y = {} & C_0 + C_1X_1 + C_2X_2 + C_3X_3 + C_4X_4 + C_5X_{12} \\ & + C_6X_{22} + C_7X_{32} + C_8X_{42} + C_9X_1X_2 \\ & + C_{10}X_2X_3 + C_{11}X_3X_4 + C_{12}X_4X_1 \ldots \end{aligned}$$

where, C_0 is constant and C_1, C_2 … C_9 are coefficient of the regression equation, X_1, X_2, X_3, and X_4 are the parameters, i.e lycra (denier), stitch length (mm), stretch (%), and hardness of limb (degree).

3 Results

3.1 Effect of Lycra Denier on Pressure Behaviour of Tubular Bandage

In order to analyse the effect of lycra denier on pressure, three different deniers (20, 30 and 40 denier) were used for the cotton/lycra tubular knitted fabric production. It was observed that in plain knitted fabrics, when the lycra denier increases, there is a significant increase in pressure values as shown in Fig. 1. This increase in pressure can be explained on the basis of retraction force developed by bandage. As the lycra

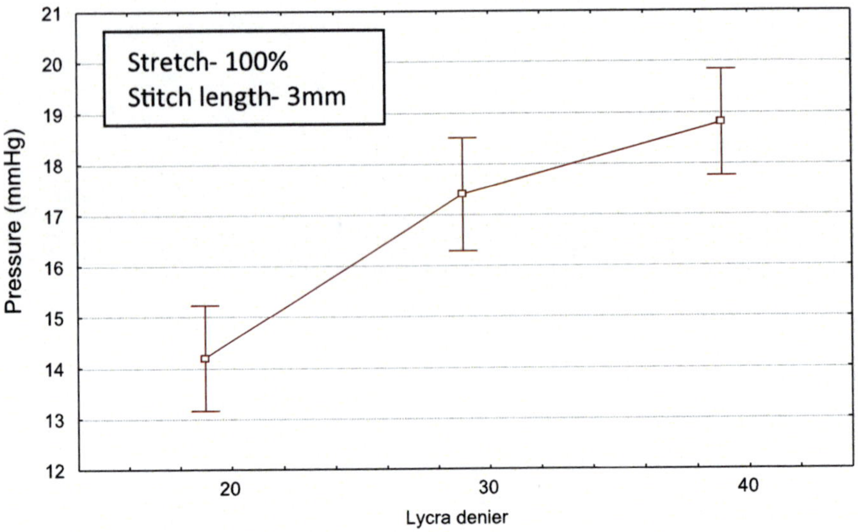

Fig. 1 Effect of lycra denier on pressure behaviour of tubular bandage

denier increases, lycra content in yarn and the weight per unit length of yarn increases which causes the increase in resulting retraction force inside the yarn. Such increase in retraction force offered by each yarn results in the overall increase in retraction force offered by bandage as the pressure offered by bandage is directly related with the retraction force developed by it. Therefore, bandage made of higher lycra denier yarn results in higher pressure generation. The increase in retraction force inside bandage can be explained through load extension graph of such bandages as shown in Fig. 2. From the graph it can be seen that at any extension, bandage with higher lycra denier yarn shows higher load value as compare to bandage with lower lycra denier. Therefore, such higher load in higher lycra denier bandage provides higher retraction force and leave higher pressure.

3.2 *Effect of Stitch Length on Pressure Behaviour of Tubular Bandage*

As per ANOVA analysis Table 2, it was found that the stitch length is the most important factor which affects the pressure developed by tubular knitted bandage. In Fig. 3, it can be seen that as the stitch length increases the pressure developed by bandage decreases significantly. Thus, the samples with the lowest stitch length exert the highest pressure. The reason for such decrease in pressure with the increase of stitch length can be explained on the basis of stretch ability and retraction characteristics of knitted fabric under tension. Stitch length of knitted fabric basically shows

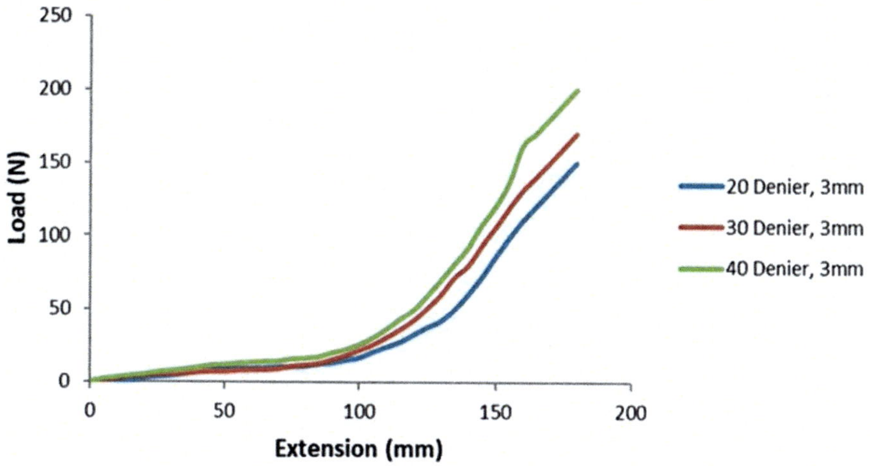

Fig. 2 Load Extension behaviour of three different lycra denier at 3 mm stitch length

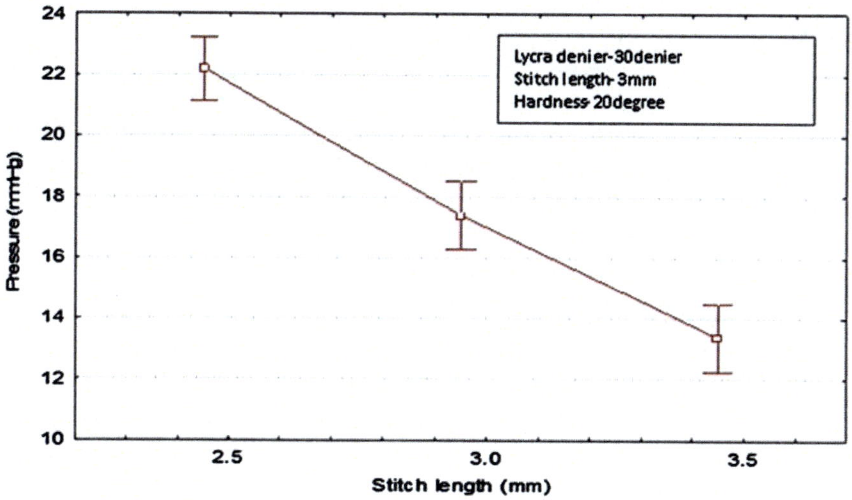

Fig. 3 Effect of Stitch length on pressure behaviour of tubular bandage

its ability to stretch. Higher the stitch length of knitted fabric higher will be its ability to stretch and therefore shows higher extensibility but the retraction force developed by fabric depends on its ability to stretch and stretch %. Higher the ability of fabric to stretch lower will be retraction force experienced by it but higher the stretch % inside fabric higher will be the retraction force. At constant stretch %, fabric with lower stitch length experience more retraction force because it opposes more during its stretch as there is no sufficient length of yarn for its extension within the fabric structure. Therefore, at constant stretch % bandage with lower stitch length results in

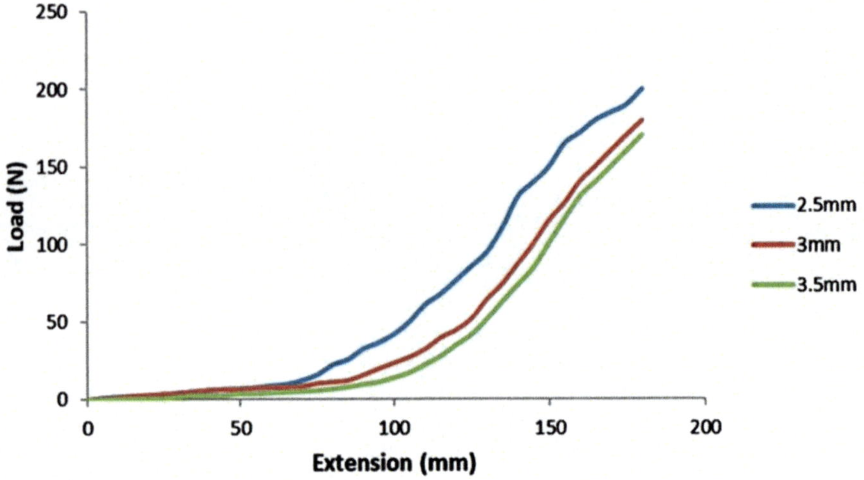

Fig. 4 Load Extension behaviour of 30 denier lycra at three stitch length

more retraction force and so develops more pressure. The increase in retraction force inside bandage can be explained through load extension graph of such bandages as shown in Fig. 4. From graph it can be seen that at any extension, bandage with lower stitch length shows higher load value as compare to bandage with higher stitch length. Therefore, such higher load in low stitch length bandage provides higher retraction force.

3.3 Effect of Stretch % on the Pressure Behaviour of Tubular Bandage

It can be observed from Fig. 5, that as the stretch % inside bandage increases the pressure developed by it increases. It can be explained on the basis of the relation between stretch % and tension developed inside bandage. Stretch inside the bandage is directly related with the tension developed by bandage. It can also be seen from load elongation graph as shown in Fig. 6, that as the elongation % increases the value of load which is basically the tension borne by bandage specimen also increases. Therefore, with the increase of stretch % tension inside bandage increases and so the resulting pressure offered by bandage also increases.

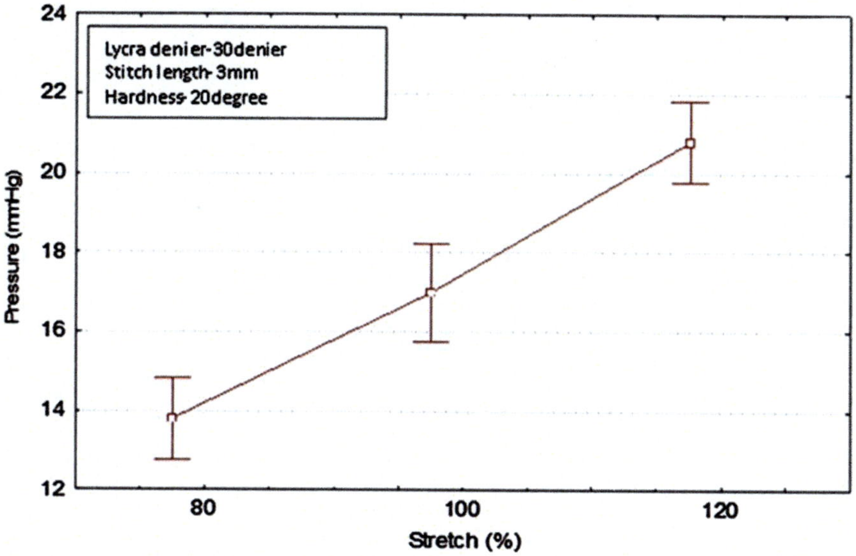

Fig. 5 Effect of Stretch % on the pressure behaviour of tubular bandage

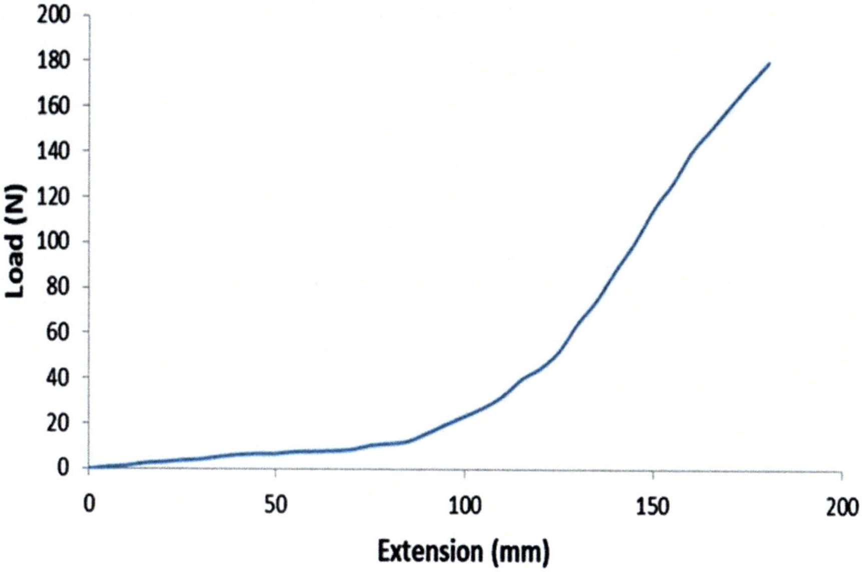

Fig. 6 Load Extension behaviour of 30 denier lycra at 3 mm stitch length

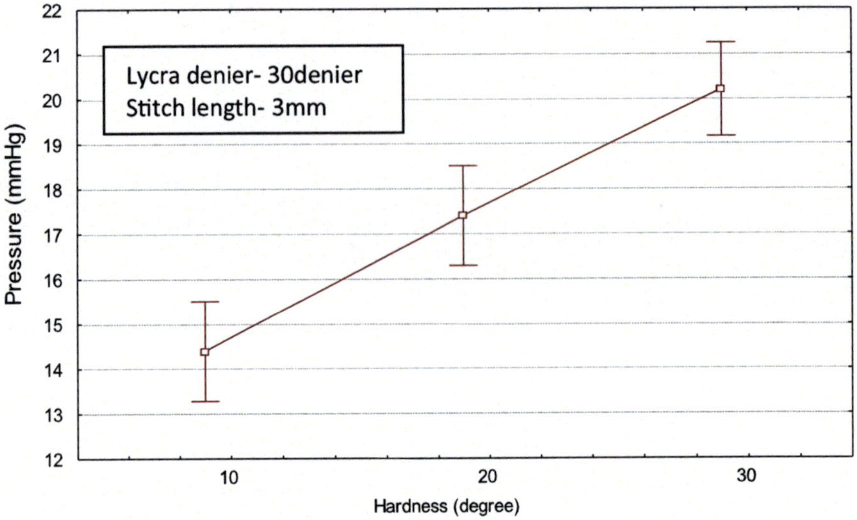

Fig. 7 Effect of limb hardness on pressure behaviour of tubular bandage

3.4 Effect of Limb Hardness on Pressure Behaviour of Tubular Bandage

It can be observed from Fig. 7, that as the hardness of limb circumference over which the bandage is to be applied increases the pressure developed by bandage also increases. The reason for such increase in pressure can be explained on the basis of reaction force experienced by bandage from the surface of limb. When the bandage is applied on harder limb the reaction force offered by the surface of limb becomes more as compare to softer limb therefore pressure sensor senses higher pressure at harder surface.

3.5 Combined Effect of Different Parameters on Tubular Bandage Pressure

From the above section it has been concluded that the change in individual parameter directly affects the pressure developed by bandage. But interaction effect of two or more parameters also has important role in contributing to pressure applied by bandage. It can be observed from ANOVA Table 2, that the interaction of different parameters also has significant importance on the pressure behaviour of the tubular bandage.

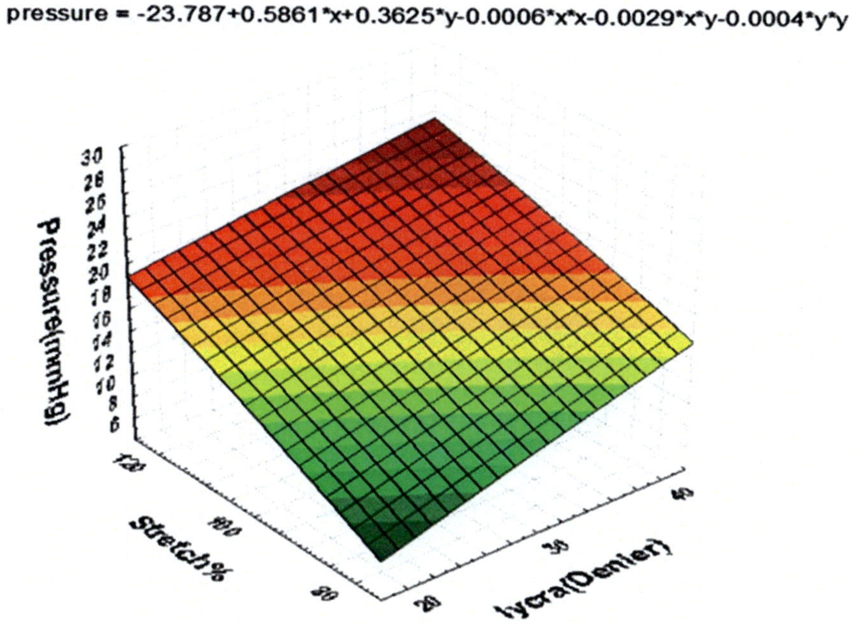

Fig. 8 Interaction effect of stretch % and lycra denier on tubular bandage pressure

3.5.1 Interaction Effect of Stretch % and Lycra Denier on Tubular Bandage Pressure

It can be observed from Fig. 8, that with the increase in stretch (%) and lycra denier, pressure developed by bandage also increases. The value of pressure is lowest at the lower level of lycra denier and stretch % while it is highest at their higher level. In between these levels the pressure varies with the equation shown in Fig. 8.

3.5.2 Interaction Effect of Stitch Length and Lycra Denier on Tubular Bandage Pressure

It can be observed from Fig. 9, that with the increase in stitch length pressure developed by bandage decreases but it increases with the increase of lycra denier. While considering their interaction effect, value of pressure is highest at the lower level of stitch length and higher level of lycra denier while it is lowest at higher level of stitch length and lower level of lycra denier. In between these levels the pressure varies with the equation shown in Fig. 9.

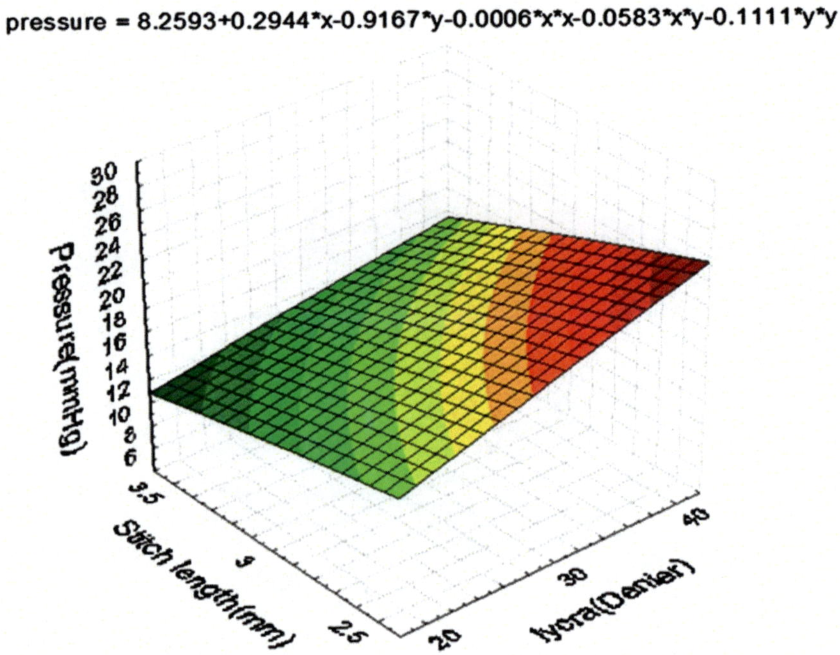

Fig. 9 Interaction effect of stitch length and lycra denier on tubular bandage pressure

3.5.3 Interaction Effect of Limb Hardness and Lycra Denier on Tubular Bandage Pressure

It can be observed from Fig. 10, that with the increase of limb hardness and lycra denier the pressure developed by bandage also increases. The value of pressure is lowest at the lower level of limb hardness and lycra denier while it is highest at their higher level. In between these levels the pressure varies with the equation shown in Fig. 10.

3.5.4 Interaction Effect of Stretch % and Stitch Length on Tubular Bandage Pressure

It can be observed from Fig. 11, that with the increase in stretch %, the pressure developed by bandage increases but it decreases with the increase of stitch length. While considering their interaction effect, value of pressure is highest at the lower level of stitch length and higher level of stretch % while it is lowest at higher level of stitch length and lower level of stretch %. In between these levels the pressure varies with the equation shown in Fig. 11.

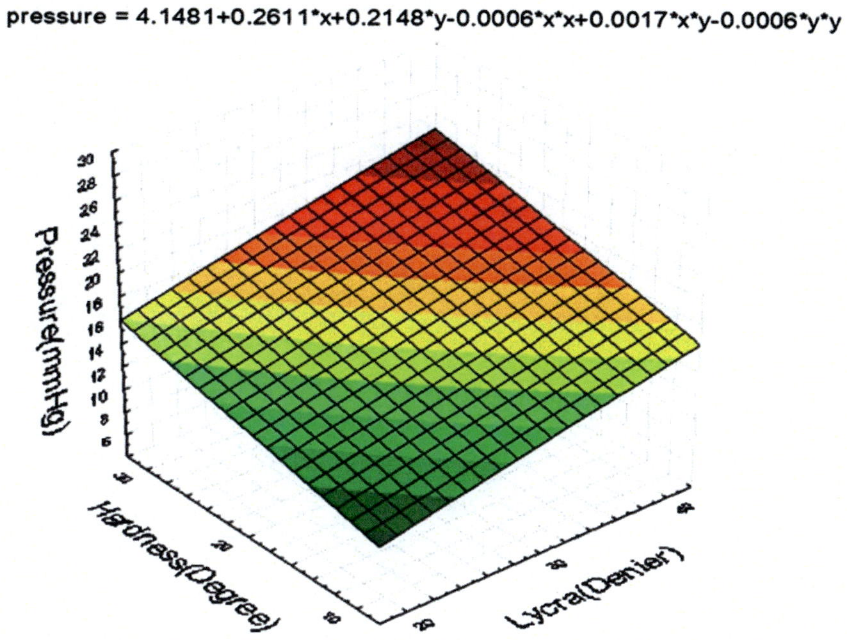

Fig. 10 Interaction effect of limb hardness and lycra denier on tubular bandage pressure

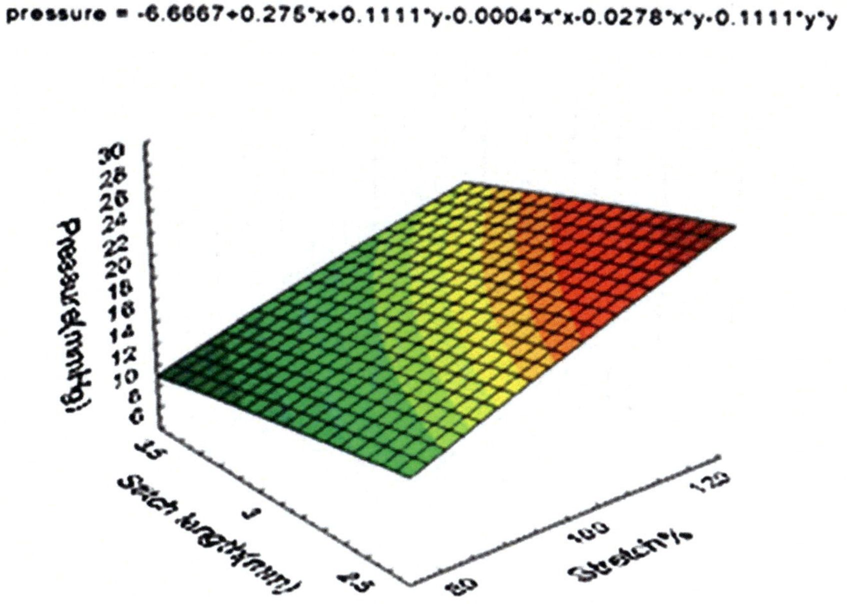

Fig. 11 Interaction effect of stretch % and stitch length on tubular bandage pressure

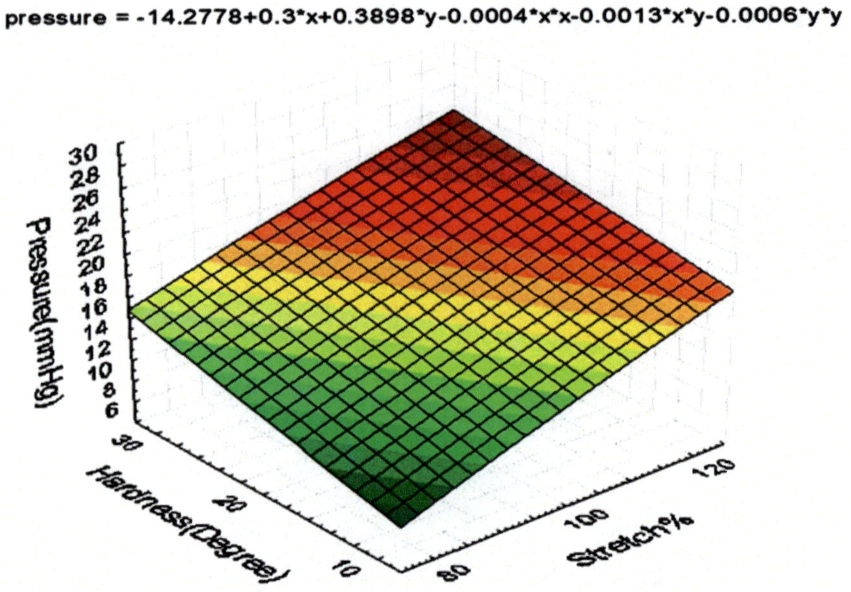

Fig. 12 Interaction effect of limb hardness and stretch % on tubular bandage pressure

3.5.5 Interaction Effect of Limb Hardness and Stretch % on Tubular Bandage Pressure

It can be observed from Fig. 12, that with the increase of limb hardness and stretch % the pressure developed by bandage increases. The value of pressure is lowest at the lower level of limb hardness and stretch % while it is highest at their higher level. In between these levels the pressure varies with the equation shown in Fig. 12.

Analysis of variance was carried out to find the importance of each parameter and their contribution in imparting overall tubular bandage pressure.

However, the contribution of each parameter on tubular bandage pressure has been calculated from Eq. (1) and shown in Table 3.

$$\text{Contribution (\%)} = \frac{[\text{SS}_p - (df * \text{MS}_e)]}{\text{TSS}} * 100 \tag{1}$$

where SS_P = Sum of square of the parameter, df = Degree of freedom of the parameter, MS_e = Mean square of pooled error, TSS = Total sum of square.

Table 3 Contribution (%) of parameters on tubular bandage pressure-ANOVA

Property	Parameters	Contribution (%)
Pressure (mmHg)	Lycra	20
	Stretch%	40
	Stitch length	20
	Hardness	16
	Lycra*stretch%	1
	Lycra*stitch length	0.8
	Stretch%*stitch length	1.1
	Lycra*hardness	0.2
	Stretch%*hardness	0.4
	Stitch length*hardness	0.3

From Table 3, it is recognised that stretch (%) has almost 40% contribution in imparting overall bandage pressure

4 Conclusion

The pressure developed by tubular bandage depends mainly on its structure and construction parameters. In the present study, all the parameters including lycra denier, stretch %, stitch length and hardness of limb has significant impact on pressure developed by tubular bandage. With the increase in the hardness of the surface of limb, the resulting pressure developed by bandage also increases. It is mainly due to more reaction force offered by the harder surface as compared to softer one. The increasing stretch % results in the pressure developed by bandage since more tension developed by the bandage at higher stretch %. As compare to other constructional parameters, stretch (%) of tubular bandage has most significant impact on bandage pressure. From ANOVA analysis it is concluded that stretch (%) has almost 40% contribution in imparting overall bandage pressure whereas stitch length, lycra denier and limb hardness has 20%, 20% and 16% contribution respectively.

References

1. Bush R, Brown K, Latz C (2015) Compression therapies for chronic venous leg ulcers: interventions and adherence. Chronic Wound Care Manag Res 11
2. Tickle J, Ovens L, Mahoney K, Hunt S, Harris E, Hodgman L (2017) A proven alternative to compression bandaging. J Wound Care 26(4):S1–S24
3. Comerota A et al (2009) Compression in venous leg ulcers: a consensus document. World Counc Enteros Ther J 29(4):20–28
4. Kumar B, Das A, Alagirusamy R (2014) Effect of material and structure of compression bandage on interface pressure variation over time. Phlebology 29(6):376–385
5. Partsch H (2007) The Vein book- mechanism and effects of compression therapy, vol 33(11). Elsevier Ltd

6. Bhuvaneswar CG, Epstein LA, Stern TA (2007) Reactions to amputation: recognition and treatment," Primary care companion to the journal of clinical psychiatry, vol 9, no 4. Physicians Postgraduate Press Inc., pp 303–308
7. Flat-Knit & Circular-Knit Garments for Lymphedemal MyLymph (2019) [Online]. Available: https://www.mylymph.com/2017/08/02/flat-circular-knit-lymphedema-garments/. Accessed 21 Nov 2019
8. Gokarneshan N (2017) Design of compression/pressure garments for diversified medical applications. Biomed J Sci Tech Res 1(3):806–813
9. Choi KF, Kim KL (2004) Fiber segment length distribution on the yarn surface in relation to yarn abrasion resistance. Text Res J 74(7):607–610
10. Eltahan E (2016) Effect of lycra percentages and loop length on the physical and mechanical properties of single jersey knitted fabrics. J Compos 2016:1–7

Zea Mays Husk Reinforced Epoxy Composites

Harwinder Singh and Arobindo Chatterjee

1 Introduction

Use of natural fillers as reinforcement in polymer matrices is encouraged nowadays in order to compensate the rising demand of synthetic polymers for making composite materials in various technical applications. The focus has been shifted from high performance engineering materials toward socio-economic and efficient utility-based products. Numerous natural agro-based lingo-cellulosic fibers such as kenaf, sisal, coir, jute, hemp, rice husk, wheat straw, bamboo, banana fiber, etc., have been used as alternative reinforcement with the polymer matrices [1–5]. Delignification of zea mays has been carried out in different ways by researchers for fiber extraction [6–10], but in current work partial delignification of ZMH is done and used as reinforcement in the epoxy matrix.

2 Materials and Methods

2.1 *Materials*

Araldite Epoxy Resin AW 106 and Hardener HV-953 IN, manufactured by Huntsman International (India) Pvt. Ltd. are used as matrix for making the composites. ZMH is procured from local market. NaOH is used in pallet form by sd-fine Chemicals Ltd. for the alkali treatment. Customized moulds are prepared for fabrication of composites from Central Institute of Hand Tools, Jalandhar, India.

H. Singh (✉) · A. Chatterjee
Department of Textile Technology, Dr. B. R., Ambedkar National Institute of Technology Jalandhar, Punjab, India
e-mail: harwinders.15.phd@nitj.ac.in

V. Midha and A. Mukhopadhyay (eds.), *Recent Trends in Traditional and Technical Textiles*, https://doi.org/10.1007/978-981-15-9995-8_6

2.2 *Alkali Pretreatment*

The pretreatment process is optimized by using, 3 factors (alkali concentration, treatment temperature and time)- 3 level, Box Behnken design of experiment. ZMH is immersed in alkali concentrations (10, 15 and 20 g/l) at 1:40 MLR solution for given time (30, 60 and 90 min) and (30, 40 and 50 °C) temperature. After the alkali treatment ZMH are washed with 5% acetic acid solution to neutralize the samples followed by drying. Fabrication of Composites is done by hand lay-up method.

2.3 *Characterization Techniques*

The XRD is done using Empyrean diffractometer made by Melvern Panalytical. The diffraction angle (2θ) is taken in the range of 5°–60°. SEM analysis is carried out by JEOL (6510LV) at 500x magnification level. DMA tests are performed using DMA1 dynamic mechanical analyzer (Mettler Toledo) in 3 point bending mode.

3 Results and Discussion

3.1 *Effect of Pretreatment on Tensile Strength of ZMH*

The effect of the pretreatment on breaking strength of the ZMH is shown in Fig. 1. Maximum breaking strength of 35.5 kPa is obtained at 15 g/l of NaOH concentration, 30 min time and at 30 °C whereas least is around 10 kPa at 20 g/l of NaOH concentration, 60 min time and at 50 °C. The reduction in breaking strength at higher levels of NaOH concentration, time and temperature is due to enhanced delignification at elevated temperature, time and NaOH concentration [11]. However, reason for lowered breaking strength in case of 10/30/40 (16.2 kPa) as compared to at 20/30/40 (27 kPa) and within 15/60/40 may be attributed to the natural variation in the thickness of ZMH which is inevitable. Defibrillation is observed at elevated temperature and time, which resulted in the loss in weight of the ZMH after pretreatment. If concentration (20 g/l) and time (60 min) are kept constant, elevated temperature (from 30 to 50 °C) is playing a vital role in further increasing the weight loss by 22% [12]. The optimized pretreatment conditions are 30 °C temperature, 20 g/l NaOH concentration and 30 min time.

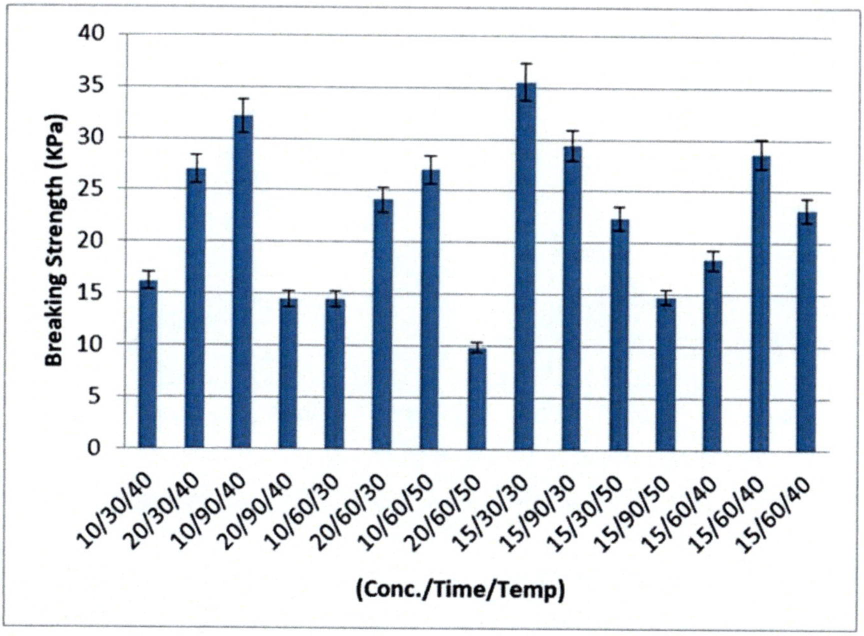

Fig. 1 Breaking strength of ZMH after pretreatment at different levels of concentration /Time/Temperature

3.2 XRD Results

Figure 2 presents the comparison of X-ray diffraction patterns of alkali treated and untreated ZMH exhibiting two main reflection peaks at $2\theta = 16.42°$ and $21.87°$. The different X-ray diffraction spectrum depends on proportion of two crystalline forms [13]. The highest peak at $2\theta = 21.87°$ corresponds to the I_{200} plane represents the both crystalline and amorphous material. The lowest peak height at $2\theta = 18.43°$ corresponds to the I_{AM} plane and represents only the amorphous part. The crystallinity of the treated sample is increased as compared to untreated one by 20.46% due to removal of lignin and partial hemicellulose.

3.3 SEM Analysis

SEM micrographs of pretreated and untreated samples are shown in Fig. 3. Array of numerous polygon type structures can be observed from Fig. 3a on the surface of untreated ZMH, while in Fig. 3b the ZMH after pretreatment appears somewhat different. There is shrinkage in the ridges (polygon shapes) and delignification of the film. Delignification is a result of the breakage of lignocellulosic bonding due to

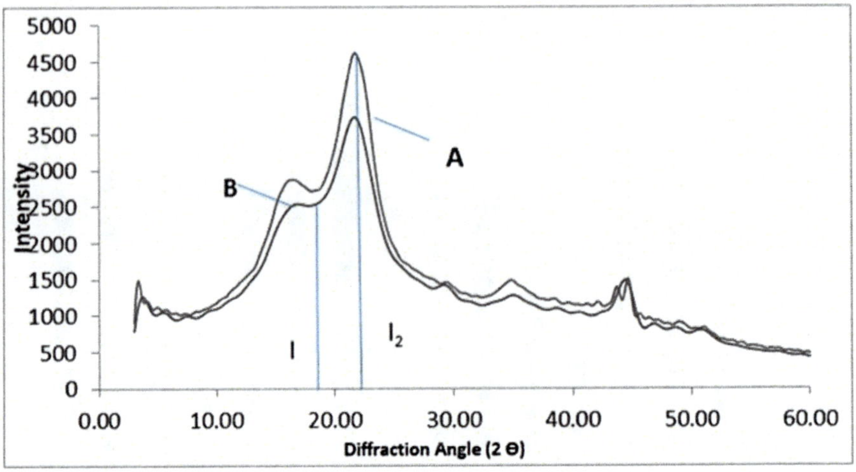

Fig. 2 XRD pattern of ZMH under varying treatment conditions viz. A. 20/30/30 B. Untreated

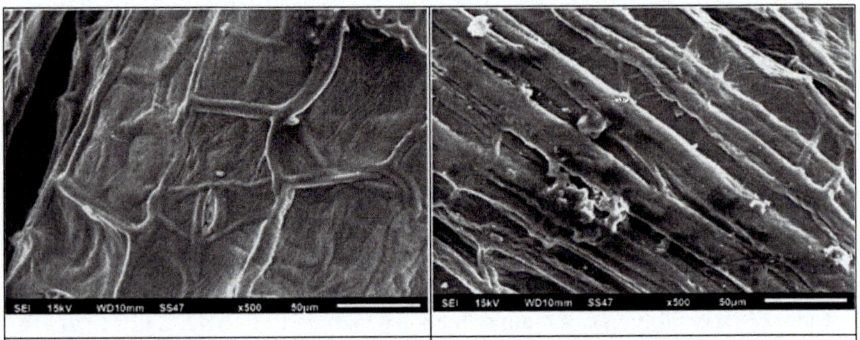

a. Untreated ZMH	b. Alkali Treated ZMH (20/30/30)

Fig. 3 SEM images of untreated and pretreated CHF

pretreatment of ZMH with NaOH. Alkali treatment causes swelling which leads to an increase in the internal surface area because of which structural linkages between lignin and cellulose are separated leading to better fiber-matrix interphase [14, 15].

3.4 Dynamic Mechanical Analysis

Figure 4 show the DMA curve of storage modulus and temperature of ZMH reinforced epoxy composite and neat epoxy. The initial modulus of ZMH reinforced epoxy composite is 3.45 GPa which is around 42% more than neat epoxy. There is a marked decrease in storage modulus of ZMH reinforced epoxy composite from

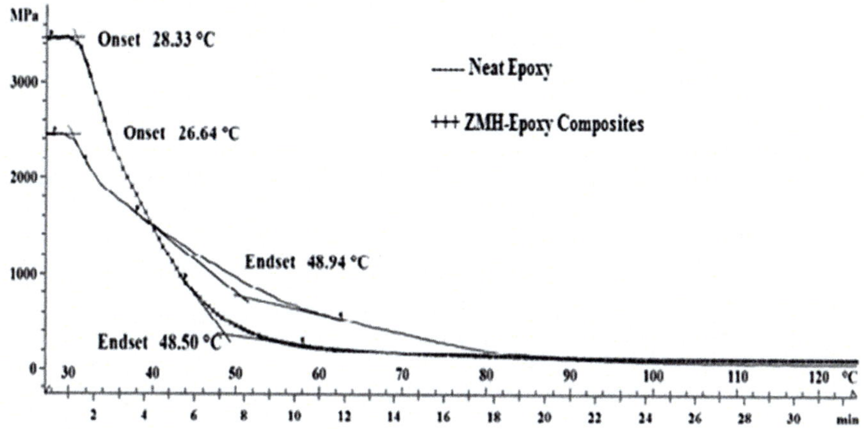

Fig. 4 Storage modulus of neat epoxy ZMH reinforced epoxy composite

3.45 GPa at 27 °C to 0.34 GPa at 48 °C, i.e. 90% decrease in storage modulus. This decrease is attributed to the softening of the polymer due to the increase in the chain mobility of the polymer matrix at high temperatures. Therefore the load bearing capacity is drastically reduced at elevated temperature [15]. Similar transition has been observed for neat epoxy but in this case, the decrease in storage modulus up to 90% has been delayed and achieved at around 75 °C [16].

4 Conclusion

ZMH can be used as reinforcement for polymer matrix. Pretreatment of ZMH is optimized (20/30/30) which resulted in surface modification of the ZMH and consequently enhanced the bonding between ZMH and epoxy. XRD spectra is governing the partial removal of lignin and hemicellulose because of which crystallinity of the ZMH is enhanced. DMA results have shown that storage modulus of the treated ZMH is 42% higher than untreated one which indicates that ZMH can be used for composites applications and could be explored further.

References

1. Thilagavathi G, Pradeep E, Kannaian T, Sasikala L (2010) Development of natural fiber nonwovens for application as car interiors for noise control. J Ind Text 39(3):267–278
2. Chauhan A, Chauhan P (2013) Natural fibers reinforced advanced materials. J Chem Eng Process Technol s6
3. N Uthaman, A Majeed, Pandurangan (2006) Impact Modification of Polyoxymethylene (POM). E-Polymers, 1–9 (2006)

4. Ku H, Wang H, Pattarachaiyakoop N, Trada M (2011) A review on the tensile properties of natural fiber reinforced polymer composites. Compos Part B Eng 42(4):856–873
5. Luo Z et al (2017) Comparison of performances of corn fiber plastic composites made from different parts of corn stalk. Ind Crops Prod 95(December 2017):521–527
6. Technology P (2015) Studies on development and characterization of polypropylene based biocomposites. Mar 2015
7. Norashikin MZ, Ibrahim MZ (2009) The potential of natural waste (corn husk) for production of environmental friendly biodegradable film for seedling. World Acad Sci Eng Technol 58(10):176–180
8. Isikgor FH, Becer CR (2015) Lignocellulosic biomass: a sustainable platform for the production of bio-based chemicals and polymers. Polym Chem 6(25):4497–4559
9. Shanmugam D, Thiruchitrambalam M (2013) Static and dynamic mechanical properties of alkali treated unidirectional continuous Palmyra Palm Leaf Stalk Fiber/jute fiber reinforced hybrid polyester composites. Mater Des 50:533–542
10. Wada M, Okano T (2001) Localization of Iα and Iβ phases in algal cellulose revealed by acid treatments. Cellulose 8(3):183–188
11. Himadri Das DK (2014) Recent development of fiber reinforced composite materials, vol 37(July) (2014)
12. Kumar R, Mago G, Balan V, Wyman CE (2009) Physical and chemical characterizations of corn stover and poplar solids resulting from leading pretreatment technologies. Bioresour Technol 100(17):3948–3962
13. Park S, Baker JO, Himmel ME, Parilla PA, Johnson DK (2010) Cellulose crystallinity index: Measurement techniques and their impact on interpreting cellulase performance. Biotechnol Biofuels 3:1–10
14. Druzian JI (2015) Extraction and characterization of cellulose. 49(2):127–133
15. Nuruddin M et al (2011) Extraction and characterization of cellulose microfibrils from agricultural wastes in an integrated biorefinery initiative. Cellul Chem Technol 45(5–6):347–354
16. Kocak D, Merdan N, Evren OB (2015) Research into the specifications of woven composites obtained from raffia fibers pretreated using the ecological method. Text Res J 85(3):302–315

A Novel Approach in Dyeing of Cotton Fabric Without Salt

M. Senthilkumar

1 Introduction

Textile dyeing is a complicated process involving the addition of a large number of dyes and auxiliaries to achieve the exact required shade. The wastes generated in this process are huge and polluting to the land and the surrounding environment. It is very toxic and greatly affects the water bodies if diverted into it. This has led to the closure of a number of textile processing industries in and around the state. The main cause of this effluent problem happens to be the addition of salt in reactive dyeing. Hence rather than finding a method to dispose these hazardous waste generated it would be better to devise a safer system of dyeing which happens to be the objective of this work [1, 2]. Salt is added in reactive dyeing as a source of positive charges to the dye bath which is inevitable in the normal dyeing method. But the positive charges can be created both chemically and electrically. The intention of the research is to replace the chemical method by electrical method. Electric current of a particular voltage is passed through the dye bath to aid in the transfer of dye molecules from the dye bath to the fabric [3–5]. With this new method a number of trials were conducted to match the shade with standard shade. The various problems encountered during the process were rectified and a perfect shade match has been achieved [6]. This has a good scope of implementation in all industries by slight modification of the present working style. This also poses a chance for reuse of dye liquor which will minimize the consumption of dyes and reduce cost. This when implemented in the industries serves a dual purpose of reducing cost and minimizing effluents thus aiding in a green dyeing process [7, 8].

M. Senthilkumar (✉)
Department of Textile Technology, PSG College of Technology, Coimbatore 641004, Tamilnadu, India
e-mail: mskumar77@gmail.com

V. Midha and A. Mukhopadhyay (eds.), *Recent Trends in Traditional and Technical Textiles*, https://doi.org/10.1007/978-981-15-9995-8_7

Table 1 Particulars of cotton fabric

S. No.	Particulars	Value
1	EPI	92
2	PPI	88
3	Warp Count	2/80 s
4	Weft Count	60 s
5	Cover Factor	22
6	GSM	150

2 Materials and Methods

2.1 Materials

2.1.1 Cotton Fabric

Bleached cotton fabric was used for this work. Cotton was chosen since it was most widely used. The particulars of the fabric are given in Table 1.

2.1.2 Dyes and Auxiliaries

Reactive dye was used for dyeing since cotton was best dyed with it. Two types of reactive dyes were used for the project. They are the cold brand reactive dye and the hot brand reactive dye. Cold brand reactive dyes are dyed with cotton fabric at room temperature. Hot brand dyes have only one reactive functional group and are dyed at an increased temperature. The dyes used for the research are given in Table 2.

Common salt (sodium chloride) was added in appropriate amounts to aid in the exhaustion stage of dyeing in the conventional method. An alkali (Sodium Carbonate) was used for fixation of the dye molecules onto the fabric in both conventional and developed methods. Distilled water was used as a dyeing medium in both the cases. Soap was used to finally wash off the surface molecules.

Table 2 CI Names of dyes

S. No.	Colour	Name	CI name
1	Red	Red HE 3B	Reactive Red 20 A
2	Yellow	Yellow HE 4G	Reactive Yellow 21
3	Blue	Blue HRL	Reactive Blue 19

Table 3 Particulars of dyeing with cold brand reactive dye

Particulars	
Dye	5%
Sodium Chloride	40 gpl
Sodium carbonate	25 gpl
Temperature	Room temperature
Time	60 min + 30 min

2.2 *Methods*

2.2.1 Production of Standard Sample

Cotton fabric was first dyed with reactive dyes using conventional method. This acted as the standard sample with which the samples from the electrical method were compared. Samples were prepared separately using cold brand and hot brand reactive dyes.

2.2.2 Dyeing with Cold Brand Reactive Dye

The recipe shown in Table 3 was followed for dyeing of cotton fabric with cold brand reactive dyes.

First a sample of pure cotton fabric was taken and weighed accurately. A dye bath was prepared taking the appropriate quantity of liquor as per the recipe. The dye was quantitatively weighed and added and then salt was added for exhaustion to take place. The material was wetted and then entered and worked thoroughly to obtain uniform dyeing. This was continued for a period of time and then sodium carbonate was added for fixation.

2.2.3 Dyeing with Hot Brand Reactive Dyes

The recipe shown in Table 4 was followed for dyeing cotton fabric with hot brand reactive dyes.

Table 4 Particulars of dyeing with hot brand reactive dyes

Particulars	
Dye shade	1, 3, 5(% ohm)
Sodium Chloride	20, 30, 40 gpl
Sodium Carbonate	15, 20, 25 gpl
Temperature	80 °C
Time	60 min + 30 min

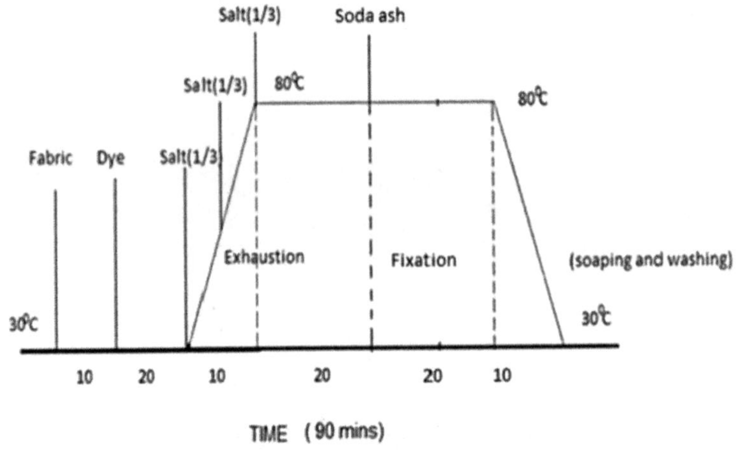

Fig. 1 Dyeing cycle of reactive dye

First a sample of pure cotton fabric was taken and weighed accurately. A dye bath was prepared taking the appropriate quantity of liquor as per the recipe. It was then placed in a bath and heated to 80 °C. This temperature was maintained throughout the process. The dye was quantitatively weighed and added and then salt was added for exhaustion to take place. The material was wetted and then entered and worked thoroughly to obtain uniform dyeing. This was continued for a fixed period of time and then sodium carbonate was added for fixation. The dyeing cycle is shown in Fig. 1.

After the dyeing was completed, washing must be done to the material to remove the extra unfixed dye molecules from the surface. This was done by means of cold wash, hot wash and soap wash. The sample was then dried thoroughly.

2.2.4 Dyeing of Cotton Fabric Using Developed Method

In this proposed method, salt was completely eliminated and electric current was used as a source of charges. The positive charge provided by the electric current on the fabric attracts the negatively charged dye molecules towards it. The following procedure was followed for dyeing.

Cold Brand Reactive Dye

The fabric sample made of pure cotton was taken and weighed accurately. A glass beaker was taken and appropriate quantity of water was added to make up the dye bath. Dye was then quantitatively added as per the recipe. This was maintained at room temperature. Two rollers made of pure stainless steel metal were designed to act as electrodes. These rollers were connected to the positive and negative terminal of the DC supply. The fabric sample was wetted and wound over the positive roller. Both the rollers were then immersed into the dye bath. Care was taken to see that the rollers did not touch each other or the walls of the container to avoid short circuit. Power was switched on and the set-up was left intact for a fixed period of time (Fig. 2). Then the fabric was unwound from the roller and fixation was done by adding alkali into the dye bath. After fixation, the sample was washed and dried.

Hot Brand Reactive Dye

The pure cotton fabric was taken and weighed accurately. The dye bath was prepared with water heated to 80 °C initially. Dye was then quantitatively added and the steel rollers were immersed into it. Fabric was wound on the positive roller. Electric current helped to maintain the temperature throughout the process. The same precautions as mentioned previously were followed and dyeing was carried out. After exhaustion with the current, fixation and washing off were done. The particulars in dyeing using the developed method are shown in Table 5.

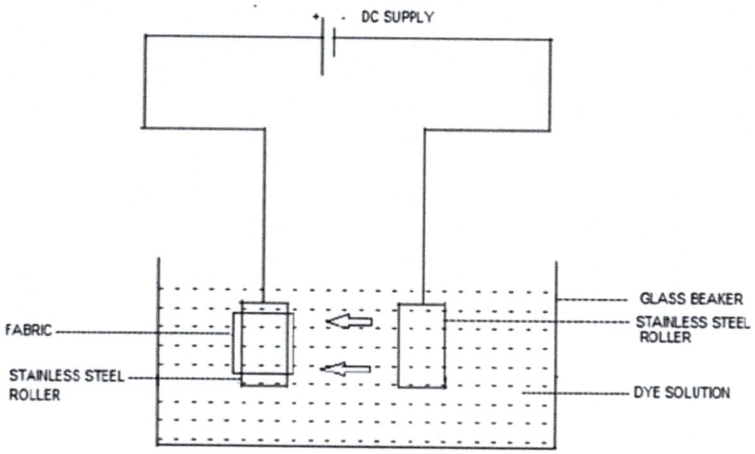

Fig. 2 Schematic representation of the developed method

Table 5 Particulars of dyeing in developed method

Particulars	
Voltage	10 V
Current	0.5—0.8 Amp
Electrode	Roller form
Metal used	Stainless Steel
Dye Bath	Glass beaker

2.2.5 Testing

Once the samples were produced from both the methods, their colour was quantitatively evaluated. It was done by using visible spectrophotometer. All the fabric samples were measured using SS5100A Spectrophotometer, Premier Colour Scan, at a wavelength range of 400–700 nm.

Measurement of Concentration of Dye Liquor

To know the percentage exhaustion of the dye onto the fabric and for various other purposes, it was important to measure the concentration of the dye liquor. This was possible in the spectrophotometer with the help of Beer Lambert's law. The law states that, The colourant concentration(C) of a solution is directly proportional to the absorbency(A) of the liquor as shown in Eq. 1.

$$A \alpha C$$
$$A = \epsilon C \tag{1}$$

where $\acute{\varepsilon}$ is a constant for a particular dye irrespective of the shade. First a dye liquor of known concentration was prepared and its absorbency was measured with the help of the spectrophotometer. This was substituted in the above equation and the $\acute{\varepsilon}$ value of the dye was found out. With this value, the concentration of the unknown liquor can be found out.

The initial concentration of the dye liquor prepared was known. After dyeing, the concentration of the residual liquor was measured by adopting the principle mentioned above. With the initial and final values of concentration, the percentage exhaustion was calculated by the formula (2) mentioned below.

$$\%\text{Exhaustion} = -\frac{\text{Initial concentration} - \text{Final concentration}}{\text{Initial concentration}} \times 100 \tag{2}$$

Dyeing Using Dye Waste Water

One of the main advantages of this method was that it allows the dye liquor to be reused again and again. In the conventional method, once salt and alkali were added, the dye liquor becomes unfit for reuse due to hydrolysation. But in this proposed method, the dye bath remains a pure dye solution since there was no salt added to it. And fixation was also proposed to be conducted in a separate bath containing alkali solution.

Using the formula given in Eq. 2, the concentration of the dyed liquor can be found out. Then keeping the initial concentration in mind, additional dye molecules are added to make it up to that level. This liquor is used for dyeing and the process was carried out three times.

Determination of K/S Value

The values of the Kubelka Munk function (K/S) of the fabrics is calculated using the below given formula (3).

$$K/S = (1 - R)^2/2R \tag{3}$$

where

R - Reflectance value
K - Absorption coefficient
S - Scattering coefficient

Reflectance value (R) of the dyed sample was measured by the Spectrophotometer.

Determination of Colour Difference Value

The values which were used to quantify a colour were the L, a, b values. L denotes the lightness or darkness value on a scale of 0 to 100. The value a denotes the greenness or redness value. A negative value denotes greener shade and a positive value denotes a reddish shade. The value b denotes blue or yellowness values. A negative value indicates a bluer shade and a positive value indicates yellow shade.

The colour difference ΔE value of the fabric was calculated using the formula (4) given below.

$$\Delta E = \sqrt{[(\Delta L)^2 + (\Delta a)^2 + (\Delta b)^2]} \tag{4}$$

where

$\Delta L =$ L_{sample}—$L_{standard}$,
$\Delta a =$ a_{sample}—$a_{standard}$,
$\Delta b =$ b_{sample}—$b_{standard}$

Measurement of Total Dissolved Solids Content

The toxicity of textile effluents was assessed by a number of parameters like BOD, COD, TDS, TSS, etc. The TDS content is especially very harmful because it cannot be removed by any means. But a salt-free dyeing significantly reduces the TDS content. It was quantitatively measured of the effluent generated from the conventional method and the developed method with the help of an instrument. The instrument used for measuring the TDS content is the Multi Parameter Water Analyser, Eutech 3210.

Table 6 Effect of voltage on colour difference values of fabrics dyed in conventional and developed method

Sample Number	Dye	Voltage (V)	ΔE Values
1	Cold Reactive yellow	60	15.173
2		50	10.55
3		40	9.289
4		30	7.712
5		20	4.624
6		10	2.324

3 Results and Discussion

3.1 Effect of Voltage on Dye Uptake of Fabric Dyed with Cold Brand Reactive Dye

The samples dyed in the proposed method needs to be compared with the samples dyed in the conventional method to ascertain that they are of the same shade. The sample dyed using the conventional method is treated as the standard against which the samples dyed using the developed method are compared with. This quantitative evaluation of the colour was done using the spectrophotometer and the results are given in Table 6 and Fig. 3. Cold brand reactive dye can be dyed at room temperature without the application of external heat. Thus, it was chosen for the initial trial works. Optimization of the process parameters was the main goal in the initial trial works while trying to achieve a perfect shade match. Voltage required for the process was varied in the range of 60–10 V and the optimum voltage was found.

Salt, in conventional dyeing, plays the role of masking the zeta potential by providing positive charges. In the proposed method, current performs this role. The effect of current on the exhaustion properties is evident from the colour achieved on the fabric. In this way, on comparing the samples dyed, it is ascertained that the exhaustion property of current is the same as that through the conventional method. Also, uniform results are achieved irrespective of the shade of the dye.

3.2 Effect of Current on Dye Uptake of Fabric Dyed with Hot Brand Reactive Dye

A reactive dye by name—Red HE 3B and CI name—Reactive Red 20 was used for trial works with hot brand reactive dye. Hot brand dyes posed an additional problem. Heat had to be provided throughout the process. But since electric current gives rise to heat, a solution was adopted. Water initially heated to 80 °C was used for dyeing and the current helped to maintain that heat throughout the process. After the process was perfected, a number of samples were dyed in the same way in the same shade

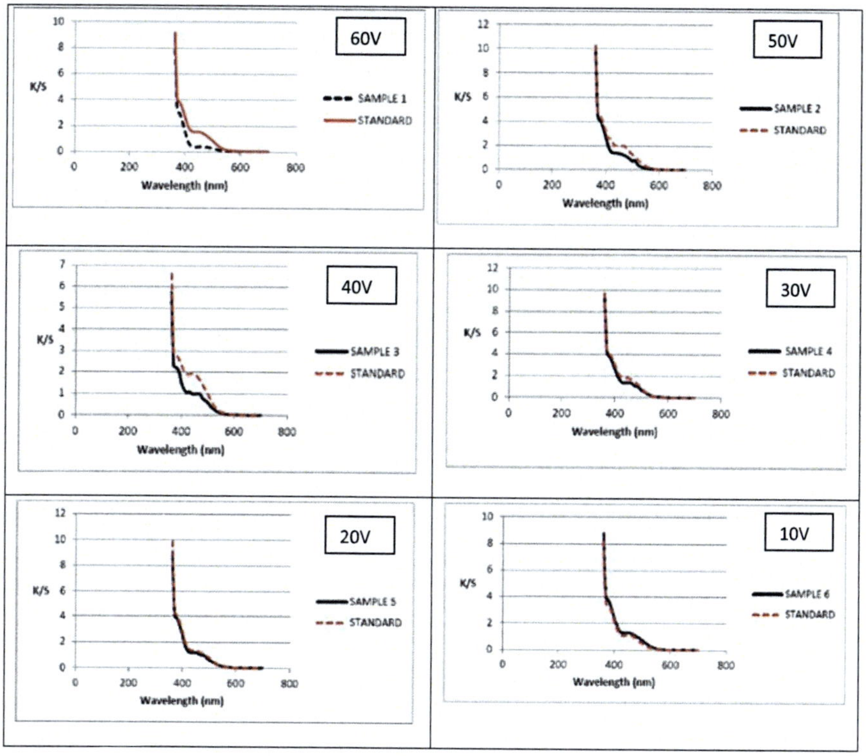

Fig. 3 Effect of current density (60–10 V) on *K*/*S* value of fabrics dyed in conventional and developed method

to check the reproducibility of the methodology. The graphs obtained between the standard samples and the ones dyed using the developed method are shown in Figs. 4, 5 and 6.

3.3 *Recycling of Dye Liquor*

Since salt is not added in the developed method, the dye liquor remains pure throughout the process. Alkali solution is also prepared in a separate bath for fixation to take place. Therefore, only water and dye molecules remain in the dye bath. Thus, reuse of dye liquor is facilitated.

First, dyeing was carried out in the developed method with fixation carried out in a separate bath. Then the liquor was measured and additional dye molecules are added to make up to the original concentration. Dyeing process is carried out again and successively samples are produced three times. The colour difference values obtained from the spectrophotometer are given in Table 7.

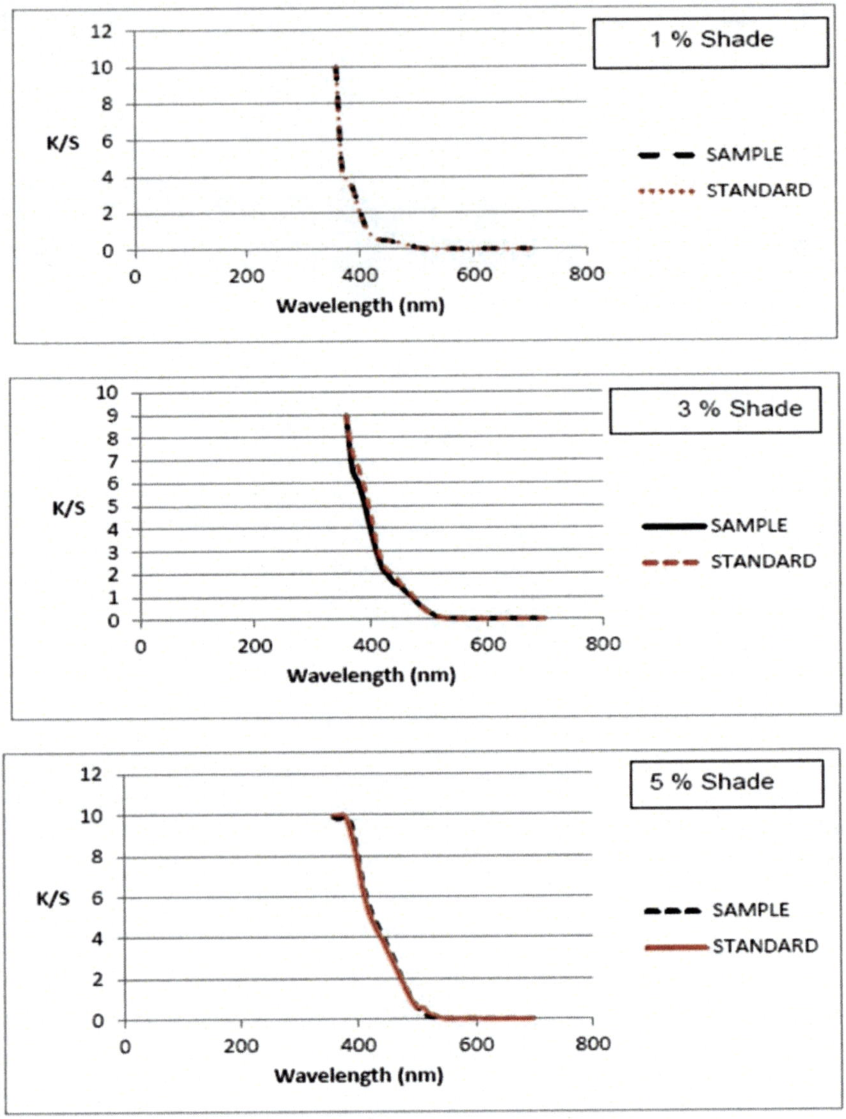

Fig. 4 *K/S* values of fabrics dyed in conventional and developed method with hot brand reactive dye (Reactive yellow 21) at 1, 3 and 5% shades

3.4 *Effect of Current on the Uniformity of the Dyed Sample*

Even deposition of the dye molecules on the fabric leads to a uniform appearance of the dyed sample. Since current flow in a conductor is uniform, the deposition of dye molecules takes place in a uniform manner. Therefore better uniformity can be

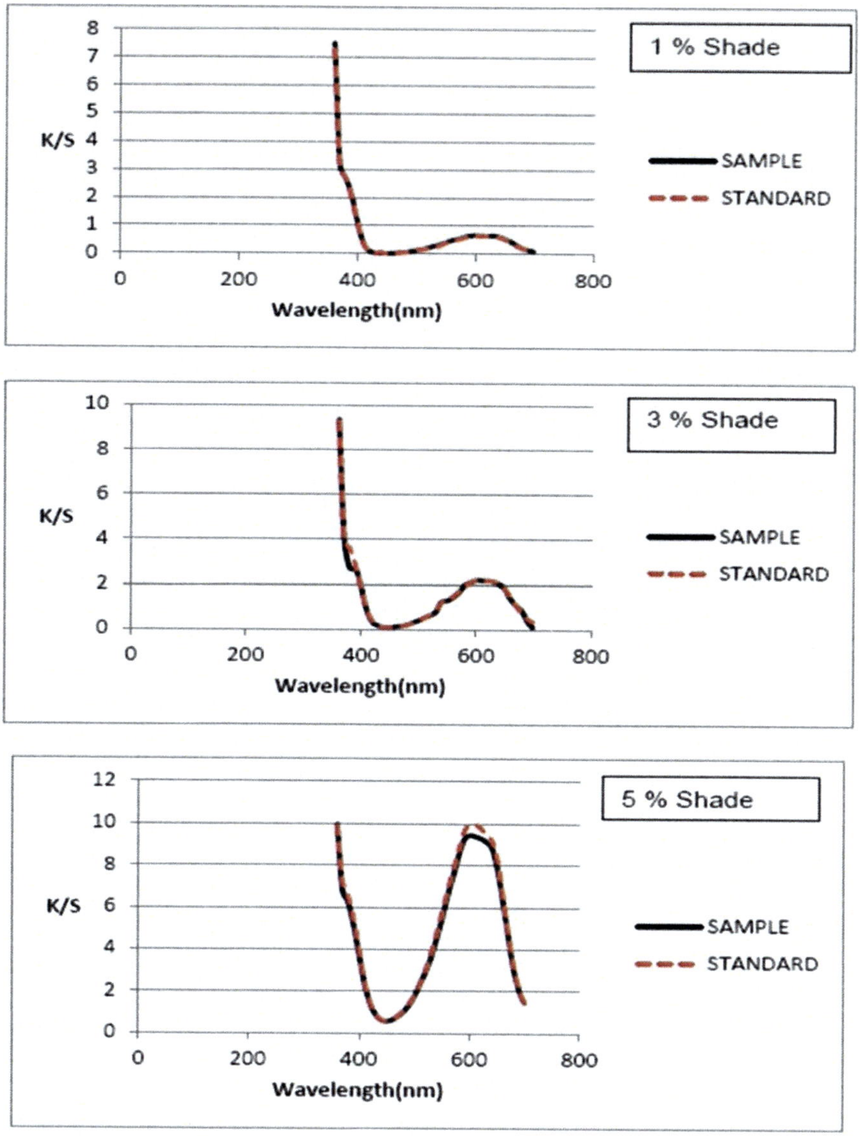

Fig. 5 *K/S* values of fabrics dyed in conventional and developed method with hot brand reactive dye (Reactive blue 19) at 1, 3 and 5% shades

achieved through the proposed method since sustaining the voltage is easier than sustaining the concentration of salt in the solution.

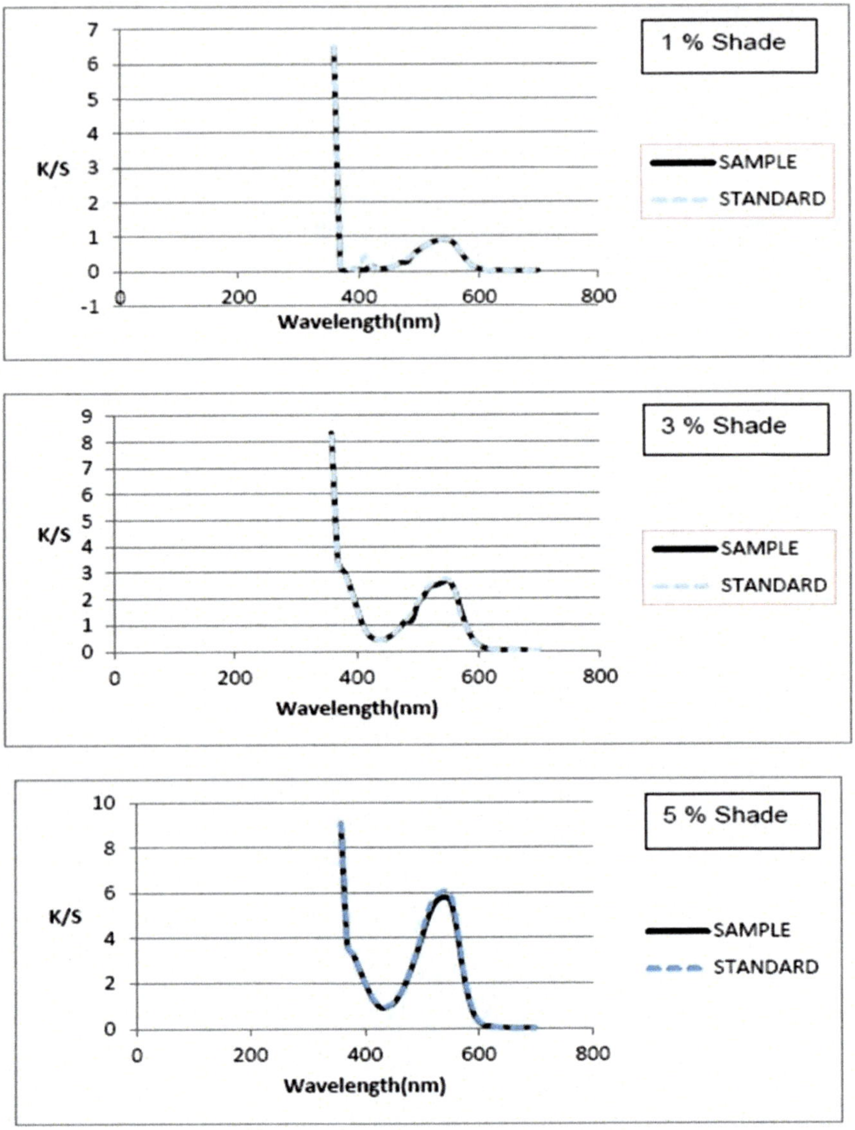

Fig. 6 *K/S* values of fabrics dyed in conventional and developed method with hot brand reactive dye (Reactive red 20 A) at 1, 3 and 5% shades

Table 7 Colour difference values of fabrics dyed from reused dye liquor

Trial	Dye	ΔE
1	Reactive Red 20 A	1.6
2		1.8
3		1.5

Table 8 Comparison of TDS values between conventional and developed method

Dye	% Shade	TDS	
		Conventional Method	Developed Method
Reactive yellow 21	1	65.65	35.42
	3	68.56	37.66
	5	63.12	32.57
Reactive Blue 19	1	62.99	31.46
	3	58.32	28.72
	5	63.47	26.95
Reactive Red 20 A	1	61.46	24.11
	3	63.25	25.98
	5	64.77	27.32

3.5 Measurement of Total Dissolved Solids Content

The TDS content is majorly contributed by the salt and it makes the effluent generated very harmful. The TDS is measured with the help of a portable TDS tester. It has probes which when inserted into the effluent solution gives a measure of the TDS directly. The TDS values obtained from both the methods are shown in Table 8. From the values given above, it is evident that the TDS values have been reduced to half in the proposed method. This reduces the harmfulness of the effluent.

4 Conclusions

The ideas pertaining to a new method of dyeing has been presented. Numerical values of the colour obtained through the two methods are compared and the results are shown.

- This elimination of salt can prove to a major breakthrough in the processing industry as it solves the age old problem of effluent generation.
- Rather than concentrating on costly equipments that would dispose the effluents, it would be more effective to modify the working method such that waste generation is minimum

References

1. MG Jason (2006) Trade waste minimisation through recycling and substitution. In: 69th annual water industry engineers and operators' conference, Bendigo Exhibition Centre, 5 to 7 Sept 2006
2. Thomas AY (2012) Dyeing of fibres using supercritical carbondioxide and electrophoresis. Patent number-20120047665
3. Ramesh Babu B, Parande AK, Raghu S, Prem Kumar T (2007) Cotton textile processing- waste generation and effluent treatment. J Cotton Sci. **11**/2007:141–153
4. Sala M, Carmen Gutiérrez-Bouzán M (2013) Electrochemical techniques in textile processes and wastewater treatment. Int J Photoenergy **12**/2012 (2012), Article ID 629103
5. Chattopadhyay DP, Chavan RB, Sharma JK (2007) Salt-free reactive dyeing of cotton. Int J Cloth Sci Technol 19(2/ 2007):99–108
6. Subramanian Senthil Kannan M, Gopalakrishnan M, Kumaravel S, Nithyanandan KJ (2006) Influence of cationisation of cotton on reactive dyeing. J Text. Apparel, Technol. Manage. 5(2/2006)
7. Singha K, Maity S, Singha M (2012) The salt-free dyeing on cotton: an approach to effluent free mechanism; can chitosan be a potential option?". Int J Text Sci 1(5/2012) pp. 69–77
8. Hebeish A, Hashem M, EL-Hosamy, Abass S (2006) No-salt dyeing behaviour of cationized linen fabrics. Res. J. Text. Apparel, Vol. 10 No. 2/2006

Localized Compression Clothing for Improved Sports Performance

P. Kandhavadivu and D. Gopalakrishnan

1 Introduction

Compression garments play an important role in optimizing the performance of a sports person in terms of speed, strength and stamina. Compression garments are elastic, body-moulded suits with an engineered compression gradient that can be worn as an upper, lower or full-body piece. An ideal compression garment is designed to provide localized compression, comfort, proper fit and stretch, optimize skin contact, absorb humidity and yet light and enduring and hence regarded as "the second level of human skin" by researchers [1, 2]. The use of compression clothing has become popular, as they increase power, improve recovery and enhance athletic performance in a variety of sports and also reduce the effects of delayed onset muscle soreness in the days following strenuous exercise [3].

When pressure is applied by garments, it compresses the soft tissues at a right angle. The pressure is applied in correspondance to vectors with magnitude and direction and hence, it increases the blood flow by increasing the hydrostatic pressure on tissues mechanically. The muscular tension is measured in terms of intramuscular pressure and this also is one of the important factor for controlling the blood flow. When the compression garments are used, the applied pressure on the skin increases the interstitial fluid pressure around the capillaries and consequently assists in the transport of excess fluid back into the circulation (veins). This further assist the local muscle blood flow, which leads to increased tissue oxygenation and enhanced muscle function [4].

P. Kandhavadivu (✉) · D. Gopalakrishnan
Department of Fashion Technology, PSG College of Technology, Coimbatore, Tamil Nadu, India
e-mail: vadivu67@yahoo.co.in

D. Gopalakrishnan
e-mail: dgk.fashion@psgtech.ac.in

V. Midha and A. Mukhopadhyay (eds.), *Recent Trends in Traditional and Technical Textiles*, https://doi.org/10.1007/978-981-15-9995-8_8

Compression garments are effective in reducing the swelling and inflammatory processes associated with muscle damage [5, 6]. This, perhaps, is because compression of the limb creates an external pressure gradient reducing the space available for swelling [7]. During the athletic performance, sometimes the damage in the muscle fibers triggers a chain of events. This damage results in an increased osmotic pressure within the tissue and leads to efflux of fluid from the capillaries of the muscle, into the interstitial spaces [8]. The application of compression may attenuate the change in osmotic pressure, subsequently reducing the degree of inflammation, the circulation of inflammatory markers and the sensation of pain [5]. A number of other claims that may explain improved performance and recovery have been proposed. These include reduced muscle oscillation and enhanced muscle pump function thought to reduce venous pooling, improve venous return, and enhance the removal of metabolites. As early as the beginning of the 1990s, Harman and co-workers found that by tightening the bandages on weightlifting athletes' knees, the vertical force output at the athletes' feet was significantly increased (amount: 11.4 ± 2.7 kg), while sports performance also indicated that wearing the bandage added 1RM for the athletes' squat thrust exercises [9].

The first sports-related study concerning compression equipment was performed by Berry and MacMurray in 1987. They found that the lower limbs' blood's lactic acid decreased after high-intensity exercise while wearing compression garments, which contributed to fatigue recovery [10]. Since then, scientists have begun to study the relationship between sports compression garments and fatigue. By observing the variations of the human body, they found that compression garments could affect fatigue. Most of the compression clothing for athletes are designed as light with mild pressure levels made of single layer fabric. Compression profiles of clothing are significantly influenced by the mechanical properties of their fabrics. **Elastane fibers are the most prominently used fiber in the compression garments for various applications. Elastane fibers are commonly used in compression garments in two ways; (i) wrapped by a natural or synthetic fiber, (ii) directly woven in to fabric [11]. In both the cases the fabric gets an added elastic nature. The type of yarn used in compression garments mainly depends on the applications**.

The wear comfort of sportswear is an important quality criterion that affects performance, efficiency and well-being of a sports person. For instance, an active sportsperson who wears poor breathable sportswear will experience a more rapid increase in the heart rate and rectal temperature than one who wears breathable sportswear. At normal condition the amount of energy expended by a person for routine activity is 50 watts/square meter/hour. During this kind of activity, the body sweats and the heat is dissipated by skin through the clothing. When the clothing worn next to skin is damp with excessive sweating, then the transportation of body heat and sweat will not be possible. This is due to the higher humid level, which ultimately reduces the absorbency and transportation of moisture in the fabric. So the moisture from skin cannot be evaporated to the atmosphere and the skin cannot be cooled immediately. This situation again creates higher temperature in the body and results in more sweat, hence the wearer starts feeling discomfort. To overcome this problem, the developed compression garments must be able to absorb the moisture from skin and also be able to transfer the moisture to the atmosphere. Athletes specializing in different endurance sports at various levels of performance, wear compression garments to improve their performance and facilitate recovery. Since compression clothing covers a large portion of the body surface, the clothing may contribute to increased body temperature since it represents a barrier to heat transfer and sweat evaporation. Since elevated skin and body temperature or hyperthermia may cause impaired endurance performance, there is a potential risk of decreased performance while wearing compression clothing in hot conditions during endurance exercise [12].

Lyocell fiber offers unique moisture transport ability and it absorbs 50% more moisture than cotton. Lyocell fiber has an extremely smooth surface and feels soft and pleasant on the skin. The combination of a smooth fiber surface and excellent moisture absorption creates a positive environment for healthy skin, making lyocell ideal even for anyone with sensitive skin. Presence of superfine or micro fibrils provide greater specific surface area, greater vapour transmission, lower flexing resistance, softer handle, greater crease resistance and greater fabric density and cover to the lyocell fiber. These advantages provide the yarn with more pores to transport vapour out by their superior capillary action. The higher pore density also provides better thermo regulation and hence lyocell can be used effectively for the development of high-performance sportswear. Similarly, the polyester microfiber yarn enables very dense fabrics to be produced with extended specific fabric or fiber surface area, developing more pores to transport vapour out by their superior capillary action. The higher pore density also provides better thermo regulation. Finer fibrils will provide higher specific area, better vapour transmission, softer handle, higher fabric density, better cover and lower flexing resistance. Comfort properties of polyester microfiber fabric are more in terms of wicking [13].

Based on the above concept, a localized compression garment was developed with lyocell fiber to improve the performance of the athletes by compressing their muscle activation, thereby enhancing performance of the wearer. A double layered single jersey plaited fabric with different elastane proportion is developed by differing the linear densities.

2 Materials and Methods

The Polyester microfiber yarn, Lyocell and Elastane yarns were sourced from the local outlets from Tirupur, Tamil Nadu, India. The fibers were knitted with double layered single jersey plaited fabric with different proportion by altering their linear densities as provided in Table 1. The face side of the fabric is made of Lyocell which has better moisture absorbency and the back side is made of Polyester microfiber which is hydrophilic, in such a way that these layers are distinct and separate yet integrated with the other.

Table 1 Fabric Specification

Fabric type	Composition	CPI	WPI	GSM	Thickness (mm)
Lower elasticity fabric	Polyester micro fiber-75 Denier, Lyocell-65 Ne, Elastane-20 Denier	51	34	237.0	1.65
Medium elasticity fabric	Polyester micro fiber-75 Denier Lyocell-65 Ne Elastane-40 Denier	57	39	309.0	1.89
Higher elasticity fabric	Polyester micro fiber-75 Denier Lyocell-65 Ne Elastane-70 Denier	60	41	425.0	1.95
Commercial fabric	Nylon Elastane	62	35	256.0	0.95

2.1 *Design Concept of Compression Sports Wear*

Shorts and tops having a localized compression effect for practicing a sport in which the legs are moved repetitively, in particular running or cycling, are constituted by stretchable textile pieces by assembling together three types of pieces, using a sewing technique that limits or eliminates any risk of non-uniformity of elasticity between assembled-together pieces. The upper body compression garment without sleeve has its compression zone in the chest and abdomen as shown in the Fig. 1. This style of garment is mostly worn by athletes. In the Fig. 1 below, the shaded area is constructed using medium elasticity fabric and the unshaded area is constructed using low elasticity fabric.

The lower body compression garment shorts are from waist to above knee length. The shorts have its compression zones in hip, thigh and buttocks as shown in the

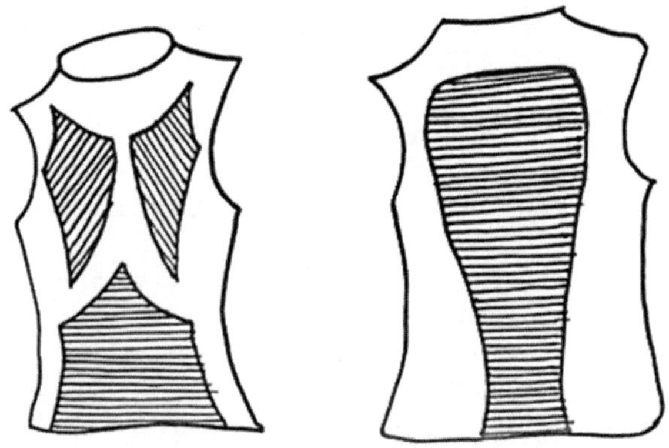

Fig. 1 Front and back of upper body garment without sleeve

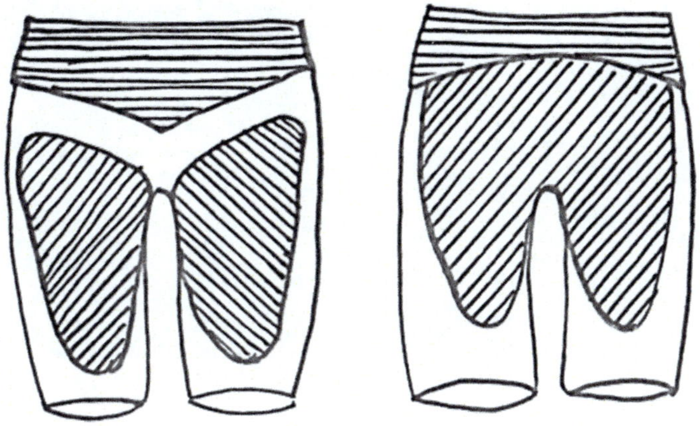

Fig. 2 Front and back of shorts

Fig. 2. In this Fig. 2, the shaded area is constructed using high elasticity fabric and the unshaded area is constructed using low elasticity fabric.

2.2 *Preparation and Physical Property Analysis*

The measurements for the compression garment is 65% of the original measurement, cut and constructed using class 500-overedge stitch and finished with the class 300–lockstitch. The double layer fabrics developed with different elastic proportions are analyzed for pressure exerted, air permeability (ASTM D737–2008), water vapour permeability (ASTM E96-80), bursting strength (ASTM D 3786), abrasion resistance (ASTM D 4966), tensile strength (ASTM D5034–2006), elongation (ASTM D 5035), elastic hysteresis (ASTM D 4964), thermal conductivity as per Lee's disk method (ASTM D 7340), moisture content and drying time (AATCC 199–2011), wickability (AATCC 197–2011), spreading rate (AATCC 198–2011) and Electromyography.

2.3 *Pressure Evaluation*

Pressure exerted on different parts of the body was measured using Kikuhime pressure sensor with self-inflatable balloon, which is capable of measuring pressure up to 200 mmHg. Electromyography was used to measure muscle activation via electric potential, referred to as (EMG), which was traditionally used for medical research and diagnosis of neuromuscular disorders (Fig. 3).

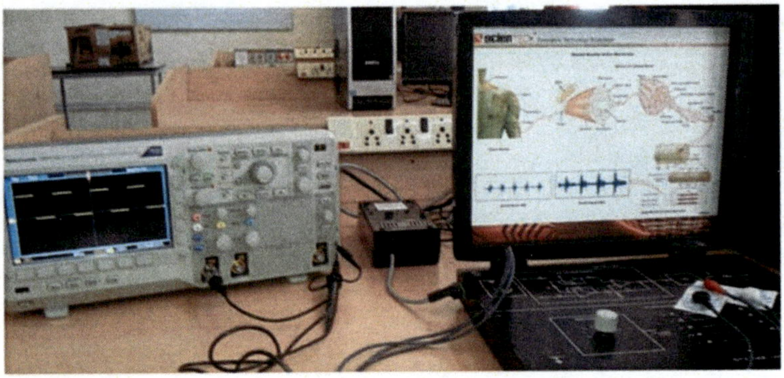

Fig. 3 Electromyography

2.4 *Field Trial Analysis*

The main objective of the field trial is to compare the muscle movement of the volunteer wearing the developed compression garment and commercial sportswear. Muscle activation was investigated using an EMG, with and without wearing localized compression garments during running. In order to standardize the running process, the mean normalized running velocity was maintained at around 5–7 m/s per minute.

- Running trials were carried out at the same speed for both with and without wearing the developed compression garment and EMG signals were recorded.
- Major muscle activations were represented by the processed EMG signals through the entire recording.
- Interfacial pressure generated at the specific area during stretch and relaxed state is measured using Kikuhime pressure sensor.

3 Results and Discussions

The compression required at different parts of the body has been studied and the level of pressure required at those specific areas are, (i) at lower thigh—10 mmHg, (ii) Upper thigh—8 mmHg, (iii) Upperbody (abdomen and chest)—15–20 mmHg. The blend analysis of different fabric composition used in this study is represented along with their physical properties in Table 2.

The blend proportion analysis revealed that the commercial fabric had 60.4% of polyester and 31.6% of regenerated cellulose content along with 8% of elastane in it. With this aspect, the low, medium and higher elastic percentage fabrics were designed with 2, 8 and 10% elastane content. The interfacial pressure generated by different elastic fabrics ranges from 6 to 21 mmHg and meets the required level of pressure as analyzed prior. Hence the compression fabrics with different range of

Table 2 Analysis of fabric properties and blend proportion

Tests	Lower elasticity	Medium elasticity	Higher elasticity	Commercial fabric
% Polyester micro fiber	63.3	60.3	58.9	60.4
% Elastomeric fiber	2.3	8.3	10.9	8
% Regenerated cellulose	34.4	31.4	30.2	31.6
Interfacial pressure mmHg	6–10	13- 16	17–21	–
Air Resistance (Kpa s/m)	0.16	0.2	0.36	0.72
Air permeability ($cm^3/cm^2/s$)	77.84	62.28	34.59	17.29
Thermal conductivity (w/m/k)	0.0074	0.0062	0.0037	0.0054
Water vapour permeability ($g/m^2/day$)	1146	1028	1035	1065
Bursting strength (lb/in^2)	120	170	186	210
Weight loss%	2.3809	1.9601	1.8181	2.3255
Moisture content %	4.33	3.57	3.49	3.03
Spreading area (cm^2)	3.28	3.05	2.51	0.66
Drying rate in (hour)	4.2	4	3.8	2.9
Wickability (mm)	190	200	210	270
Length wise strength (Kgf)	11.98	12.92	13.49	16.65
Length wise elongation (%)	148.77	198.57	284.6	233.69
Width wise strength (Kgf)	14.85	14.23	17.54	17.98
Width wise elongation (%)	294.01	247.64	348.29	300.37

elasticity were used in specific areas of garment for providing localized compression effect.

The physical property analysis revealed that the increment in elastane content from 2 to 10% had played a major role on the fabric properties. As the percentage of the elastane increases in the structure, the structure becomes tighter. In consequences, the pores between the yarns and within the yarns got reduced. This results in an increased air resistance to the fabric. The air and water vapour permeability values of the low elastic fabric are higher than the high elastic fabric. However, the commercial fabric has less air permeability and higher resistance than all the developed fabrics. But at the same time the reduction in permeability or inter and intra space reduces the thermal conductivity values of the fabric. Higher the elastane content higher will be the fabric tightness, thickness and lower porosity and so reduced thermal conductivity of the fabric. The commercial fabric has thermal conductivity similar to medium elasticity fabric. The results of Sadak et al. are in line with the findings of our research, they mentioned that the increment in elastane percentage increases the thickness and reduces the air permeability [14].

In the case of weight loss percentage, moisture content and spreading area, the increment in the elastane percentage plays a significant contribution. The increase in elastane percentage, reduces the weight loss percentage, moisture content and spreading area. This might be due to the increased thickness and compact structure. The commercial fabric had higher weight loss similar to developed low elastane fabric and the moisture content approximately closer to higher elastic fabric. In the case of spreading rate, the developed fabrics were far superior than the commercial fabric. This is one of the important factors for motivating the capillary wicking of the fabric [15]. Drying rate depicts the rate of weight loss of a wet fabric with respect to time expressed as a percentage. It is very important parameter in the case of sportswear, as the dampness created in the fabric will lead to reduced body heat and creates tiredness [16]. As the structure became compact in the high elastic fabric with 10% elastane content, the fabric compactness increases resulting in reduced drying rate. The lowest drying rate is noted for the commercial fabric compared to all the developed fabrics.

The less inter yarn space in the compact structure leading to reduced spreading rate and increased wickability. Spreading area is a measure of the extent to which a drop of water spreads on the surface of the fabric and it is a measure of the moisture management capability of the fabric. The spreading area and the moisture evaporations have great amount of link. The reduction on the spreading area subsequently shows a reduced moisture transport. The increment in the wicking might be a consequence of the closer fabric structure. The tight structure with lesser inter and intra yarn spacing increases the traverse wicking of the material. The strength and elongation percentage of the compression fabrics increases with elastane content both in length wise and width wise directions.

3.1 Elastic Hysteresis

The elastic hysteresis of different fabric composition is shown in the Fig. 4 for lower elasticity fabric (I), medium elasticity fabric (II), higher elasticity fabric (III)

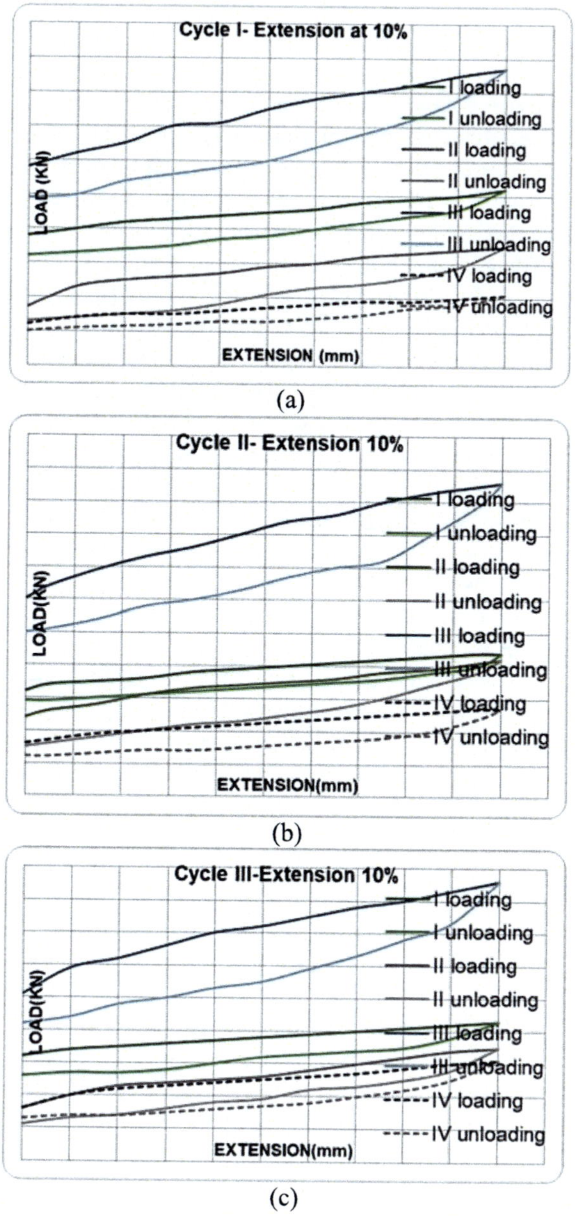

Fig. 4 Elastic Hysteresis at **a** Cycle I, **b** Cycle II and **c** Cycle III—Extension 10%

and commercial fabric (IV) respectively. From the Fig. 4, it is seen that the elastic hysteresis is very negligible and all four fabrics, for all three cycles recover to their original dimension after relaxation and thus the elastic recovery is approximately 100%.

3.2 Evaluation of Muscle Activity

Measuring muscle activation via electric potential, referred to as electromyography (EMG), has traditionally been used for medical research and diagnosis of neuromuscular disorders. Muscle activation was investigated using EMG, with and without wearing localized compression garments during running. The muscle activation is recorded as a graphical output and the amplitude in volts.

The Fig. 5a shows the muscle activation during running without compression garment. From the graph it can be observed that the muscle movement is high with the recorded amplitude of 2.00 V. The Fig. 5b shows the muscle activation while wearing lower elasticity fabric during running and the amplitude measures 500 mV. Even with low elasticity fabrics, the muscle activity seems to be reduced to great extent and the amplitude is reduced to 500 mV. The Fig. 5c shows the muscle activation during running while wearing medium elasticity fabric. The amplitude is still reduced to 200 mV. The Fig. 5d shows the muscle activation during running while wearing higher elasticity fabric and the amplitude noted was 100 mV.

By analyzing the muscle activation using EMG, it was found that the participant without wearing localized compression garment presents a larger muscle activation volume than wearing a localized compression garment. Comparing all the three elasticized fabrics, the higher and medium elasticity fabric showed 90% reduction in the muscle movement. Since, the part of the muscles' activities (muscle tuning) was proven to reduce soft tissue resonance, it is proposed that the soft tissue vibration was crucial for the energetic running. The compression garment could reduce muscle vibrations during human locomotion and these findings were in line with the previous researchers [17–19]. Hence, compression garments can be used to reduce soft tissue vibrations, prevent unnecessary muscle tuning activities and decrease energy consumption.

3.3 Field Trial

Field trial was carried out to analyze the performance of the sports person with/without wearing localized compression garment. The participant was made to run on a treadmill with a speed of 5–7 m/s and also perform cycling and the average distance covered per minute was recorded. The interfacial pressure generated during normal and stretched conditions is also recorded. As observed from literatures [20,

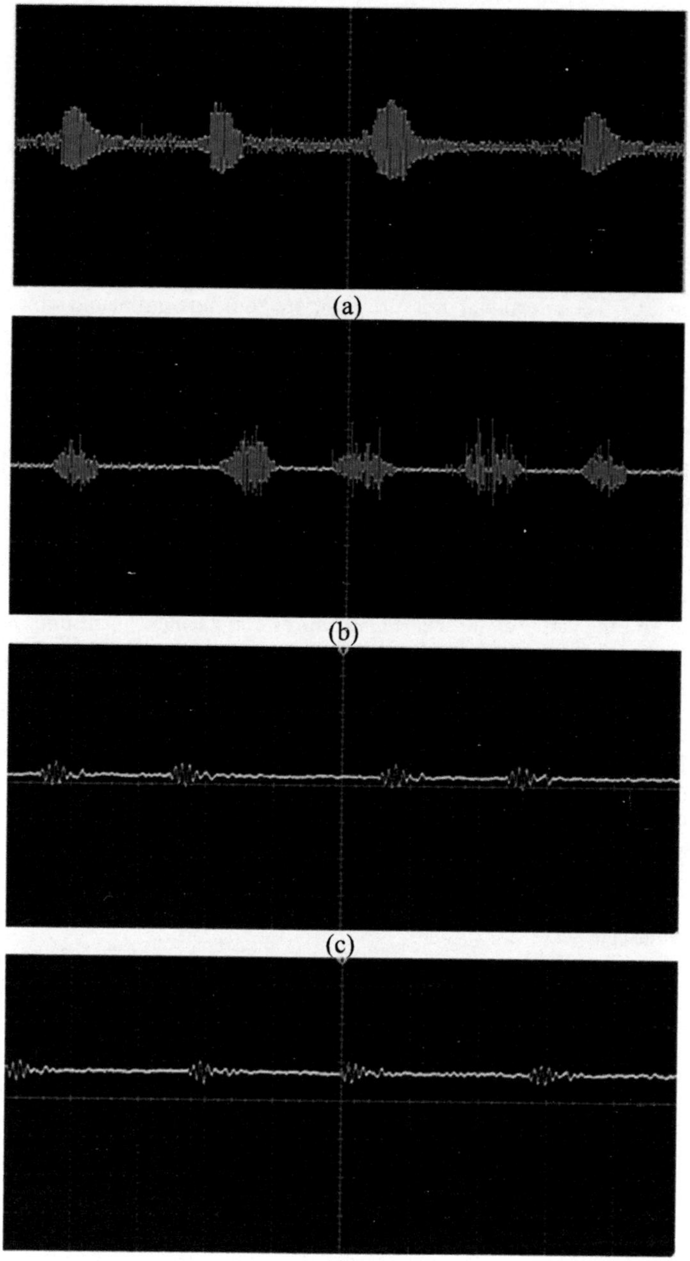

Fig. 5 **a** Muscle activation without wearing compression garment; **b** for lower elasticity fabric; **c** for medium elasticity fabric; **d** for higher elasticity fabric

21], the optimal pressure gradient which would generate fastest venous flow on specific areas of human body are shown in Table 3.

The graduated pressures generated by commercial branded sport garments ranges from 19.0 to 30.0 mmHg at the ankle to 17.6–25.0 mmHg at the calf and to 9.1–18.0 mmHg at the thigh but these claims are not substantiated by any research. The degree of pressure produced by a compression garment is determined by a complex interrelation between the following principle factors: the construction and fit of the garment; structure and physical properties of its materials; the size and shape of the part of the body to which it is applied; and the nature of the sporting activity undertaken [20]. The interfacial pressure generated during normal and stretched level while using the developed garment was measured using Kikhume pressure sensor and is listed in Table 4. The measured interfacial pressure is well within the optimum pressure and is sufficient to generate venous flow. The required level of pressure in the specific areas has been exerted by the developed localized compression garment and during field trial it was found that the participant exerted more muscle power without localized compression garment.

During field trial it is observed that the sports person could cover more distance in a specified period of time and the duration of effective activity also increased. The distance covered per minute without wearing the localized compression garment was 360 m and the same while wearing compression garment is around 395 m. An improvement of 8.8% in the performance could be observed. Similarly during cycling also the performance increased to around 10%. Compression garments provide ergogenic benefits during training and athletic performance, these proposed benefits include an increase in strength and power as well as improved endurance performance. Advantages are thought to be achieved via a number of mechanisms, which include increase in muscle oxygenation resulting from improved blood flow to the muscle and reductions in muscle oscillation thought to reduce the severity of fatigue.

Table 3 Optimal pressure level

Specific area	Optimal pressure gradient (mmHg)
Ankle	18
Calf	14
Knee	8
Lower thigh	10
Upper thigh	8
Upper body	15–20

Table 4 Interface pressure during field trial

Specific area	Normal (mmHg)	Stretch (mmHg)
Upper thigh	10	14
Under thigh	11	12
Abdomen	14	16
Chest	12	14

Use of compression garments can reduce concentrations of blood lactate when worn both during and after strenuous exercise (Fig. 6).

The experiment shows that, there is a significant increase of nearly 9% in the performance of the wearer. Hence, during running and cycling it is more advantageous to wear localized compression garment where less muscle activation is occurring thereby enhancing the performance of the wearer. If compression clothing is worn during prolonged exercise, the athletes will benefit from improved lactate elimination, reduced muscle pain, damage and inflammation during recovery. These processes are likely due to reductions of muscle oscillation during exercise, improvements in clearance of metabolites through improved blood flow, lymphatic outflow and reduced space for swelling. Potentially, this might improve recovery and enhance subsequent performance. Cycling performance seems to be more positively affected by compression clothing, A potential benefit of compression clothing is displayed during the immediate recovery from intense cycling and running since blood lactate concentrations were reduced during this period. Compression appears to exert positive effects on perceived leg or muscle soreness and delayed onset of muscle fatigue following running and cycling. The positive impact appears mostly within the following 24 h after exercise if compression clothing is worn during running.

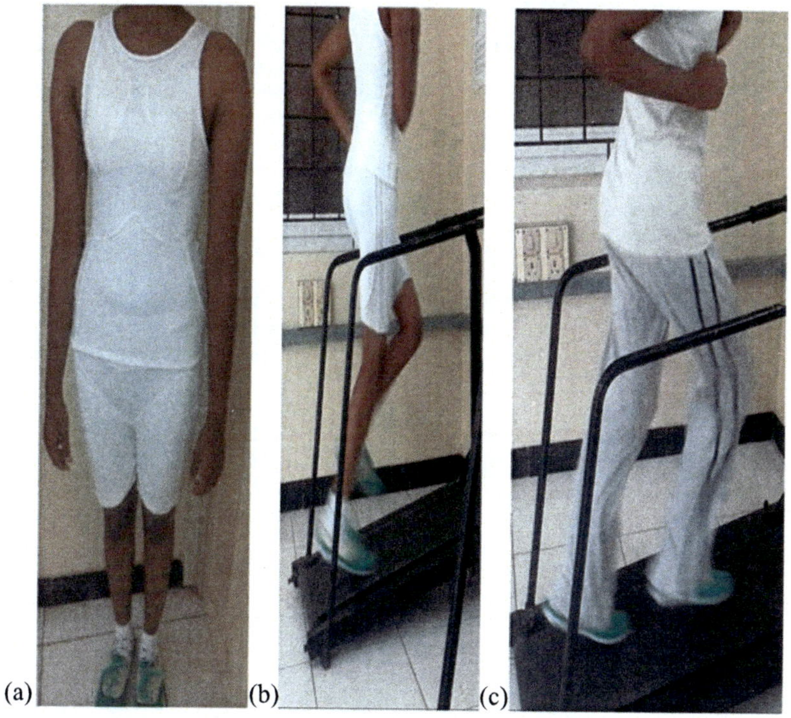

Fig. 6 Field trial images, **a** Garment appearance, **b** With developed localized compression garment, **c** With normal garment

4 Conclusion

Localized compression garment is designed and developed in such a way that, it produces localized compression effect in the specific area with the required interfacial pressure. Different composition of compression fabric made of polyester microfiber, lyocell and elastane has influence on the generated interfacial pressure. It is observed that with increase in elastane% and elastane stretch, the fabric becomes compact and thick with higher thermal resistance and reduced permeability to air and moisture vapour. As the elastane% increases, the moisture content, spreading area and drying rate are higher but the wickability is lower. During field trial, usage of medium and higher elasticity fabrics showed a reduction of 90% in the muscle movement as analyzed through EMG. The experiment resulted with 9% increase in the performance of the volunteer. Hence, it is more advantageous to wear compression garments, where less muscle activation occurs, thereby enhancing athletic performance.

References

1. Uttam D (2013) 'Active sportswear fabrics', international journal of IT. Eng Appl Sci Res (IJIEASR) 2(1):2319–4413
2. Weijie F, Liu Y, Fang Y (2013) Research advancements in humanoid compression garments in sports. Int J Advan Robot Syst https://doi.org/10.5772/54560
3. Wallace L, Slattery K, Coutts A (2006) Compression garments: do they influence athletic performance and recovery. Aust Sports Comm, Sports Coach, Online Mag Coaches 28(4):1836–2604
4. Stéphane P (2012) Compression garments: evidence for their physiological effects (P208), Eng Sport 7(2)
5. Kraemer WJ, French DN, Spiering BA (2004) Compression in the treatment of acute muscle injuries in sport. Int J Sports Med 5:200–208
6. Kraemer WJ, Bush JA, Triplett-McBride NT, Perry Koziris L, Mangino LC (1998) Compression garments: influence on muscle fatigue. J Strength Conditioning Res 12(4):211–215
7. Davies V, Thompson KG, Cooper SM (2009) The effects of compression garments on recovery. J Strength Conditioning Res 2:1786–1794
8. Kramer WJ et al (2001) Continuous compression as an effective therapeutic intervention in treating eccentric-exercise induced muscle soreness. J Sport Rehabilitation 10:11–23
9. Harman EA, Rosenstein MT, Frykman PN (1990) The effects of arms and countermovement on vertical jumping. Med Sci Sports Exerc 22(6):825–833
10. Berry MJ, McMurray RG (1987) Effects of graduated compression stockings on blood lactate following an exhaustive bout of exercise. Am J Phys Med 66:121–132
11. Senthilkumar M, Anbumani N (2011) Elastane fabrics- A tool for stretch application in sports. Indian J. Fiber Textile Res 36:300–307
12. Nybo L, González-Alonso J (2015) Critical core temperature: a hypothesis too simplistic to explain hyperthermia-induced fatigue. Scand J Med Sci Sports (Special Issue: Training and Competing in the Heat) 25(S1): 4–5, https://doi.org/10.1111/sms.12444
13. Bhat P, Bhonde HU (Nov 2006) "Comfortable clothing for defence personnel", Asian Text J 73–77
14. Sadek R, El-Hossini M, Eldeeb AS, Yassen AA (2012) Effect of Lycra extension percent on single Jersey knitted fabric properties. J Eng Fibers Fabr 7(2):11–16

15. Morent R, Geyter ND, Leys C, Vansteenkiste E, Bock JD, Philips W (2006) Measuring the wicking behavior of textiles by the combination of a horizontal wicking experiment and image processing. Rev Sci Instrum 77:093502
16. Saricam C, Kalaoglu F (2014) Investigation of the wicking and drying behavior of polyester woven fabrics. Fibres Text East. Eur 22, 3(105), 73–78
17. Born DP, Sperlich B, Holmberg HC (2013) Bringing light into the dark: effects of compression clothing on performance and recovery. Int J Sport Physiol 8:4–18
18. Duffield R, Portus M (2007) Comparison of three types of full-body compression garments on throwing and repeat-sprint performance in cricket players. Br J Sport Med 41:409–414
19. Fu W, Liu Y, Fang Y (2013) Research advancements in humanoid compression garments in sports. Int J Adv 1 Robot Syst 10, 66
20. Troynikov O, Ashayeri E, Burton M, Subic A, Alam F, Marteau S (2010) Factors influencing the effectiveness of compression garments used in sports. Elsevier Ltd, Procedia Eng 2(2):2823–2829
21. Dascombe B, Osbourne M, Humphries B, Reaburn P The physiological and performance effects of lower-body compression garments in high-performance cyclists, project report by school of health and human performance, central queensland university and centre of excellence, queensland academy of sport. https://issuu.com/skinscompression/docs/cqu_cycslist_report_original_3614 (Accessed 28 Nov 2018)

Surface Modified Hollow Polyester/Polyvinylpyrrolidone Composite for O_2 Enriched Air Separation

N. Gobi, A. Ganesh, E. Anbu, and N. Agalya

1 Introduction

Air is a natural source of various gaseous components and it is the mixture of nitrogen, oxygen, argon with suspended dust, spores, bacteria, etc. The primary gaseous as nitrogen and oxygen are needed for various industrial applications such as chemical, medical, aeronautical and food processing industry. Air enriched oxygen is useful in pyro-metallurgy (copper and nonferrous metals), combustion and furnace applications, chemical synthesis oxidation and waste water treatment [1]. Several methods are available for air separation which includes cryogenic distillation, adsorption of solid surface and solvent absorption to obtain high purity of primary gaseous components. Both of the cryogenic and adsorption techniques possess the production capability of 20 to 300 tonnes of oxygen per day and oxygen purity of more than 95% [2, 3]. Cryogenic distillation exhibits disadvantages like high energy consumption and requires high volume of vessels for air separation [4, 5].

Recently, there is a huge development in membrane for gaseous separation. It has vital application due to their inherent characteristics like simple, continuous in operation, reduced start-up time and low cost with high economical to operate. It also competes with conventional unit operations on the basis of overall economics, safety, environmental and technical aspects [6, 7].

A membrane is defined as a selective barrier between two different phases. Gas separation processes require a membrane with high permeability and selectivity and these parameters decide the performance of a membrane material [8]. Also, membrane separation of gaseous components rate is based on pressure of feed, permeate and ration of feed flow rate, membrane area and recycling. Gas separation in membranes occurs due to different levels of solubility in and diffusion through

N. Gobi (✉) · A. Ganesh · E. Anbu · N. Agalya
Department of Textile Technology, Anna University, Chennai 600025, India
e-mail: gobsnn@gmail.com

V. Midha and A. Mukhopadhyay (eds.), *Recent Trends in Traditional and Technical Textiles*, https://doi.org/10.1007/978-981-15-9995-8_9

polymeric structures. Solubility and diffusion are the thermodynamic property and kinetic quantity respectively [9]. The permeability of a gas through a membrane is closely associated with Hendry's law for simple and non-interacting gases. Relative permeability of the gaseous compound decides the degree of separation of individual components.

Solution-diffusion model has three step process; dissolution of the permeant gas into upper surface of the polymer, diffusion of the gas through the internal structure of polymer and desorption of the permeant from the lower surface of the polymer [10]. The permeability co-efficient of the particular component can be described as the product of the solubility and diffusion coefficient of the permeant [11–13]. The membrane selectivity is related with the ratio of more permeable component to less permeable component. The resistant of the molecular flow is the function of chemical and physical properties of the penetrant. As the membrane selectivity increases, the product mole fraction increases.

This mechanism is also improved by adding fillers as specific absorbers in polymeric matrix membranes to improve its selectivity. The type of fillers and proportion of fillers are selected depends on the rate of permselectivity and also the productivity of the selected permeant [14–17]. The O_2/N_2 gas pair has different diffusivity in membrane transport. In general both gases have relatively small solubility interactions with polymeric materials. According to solution-diffusion theory, the diffusivity component is thought to be more responsible for transport [18, 19].

Silicalite has significant potential for increasing the permeability and selectivity for various polymers for the separation of O_2 from N_2, CO_2 from H2 and CO_2 from CH_4.

Highest O_2/N_2 selectivity and permeability is recorded by PI carbon membrane and it is in the range of 200 to 800 Barrer and 7.5 to 15 respectively. PVP/carbon blended membrane gives higher O_2 permeability and O_2/N_2 selectivity [20]. Table 1 shows the O_2 permeability and O_2/N_2 selectivity of various polymer and its combinations.

Polymeric hollow fibres are widely used for gas separation applications and the polymeric material is permeable to oxygen and nitrogen, however the rate of oxygen permeation is several times faster than the nitrogen permeation and in general cartridge type hollow fibre membrane was used. Polysulfone hollow fibre membrane was used to enrich oxygen; the pressure and temperature influences the purity of oxygen. Various polymers like polyimide, polyaramide, cellulose acetate are used for developing hollow fibres membranes for the production of oxy-enriched air.

Table 1 Variables and levels of the experimental plan

Variables	Levels		
	−1	0	1
X_1 (Time of NaOH treatment of polyester), h	1	2	3
X_2 (PVP), (owm) %	5	10	15
X_3 (Activated carbon), (owm) %	2	4	6

In this study, a novel technique was used to fabricate the membrane such as the hollow polyester fibre is surface treated and axially arranged, and Polyvinylpyrrolidone and activated carbon are mixed and loaded in the axially arranged surface treated hollow fibre membrane. The performance on oxygen separation and other properties are analysed in detail.

2 Materials and Methods

Hollow polyester (Polyethylene terephthalate) fibre of 6 Denier and 51 mm staple length from Reliance industries was procured and analytical grade sodium hydroxide (NaOH) was used for surface treatment. Polyvinylpyrrolidone $(C_6H_9NO)_n$ (PVP) of molecular weight 30,000 was used as a thermally labile polymer and activated carbon was used for constructing the polymeric composite for air separation.

2.1 *Alkaline Hydrolysis of Polyester*

The fibres are well opened by carding machine and treated with NaOH Solution for alkaline hydrolysis, 25gpL of NaOH was taken with the material to liquor ratio of 1:25 and treatment temperature was maintained as 60 °C. The hydrolysis process was influenced by treatment time; hence, three different time duration as 1 h, 2 h and 3 h was taken for treating polyester. It was found that the further increase in treatment time as 4 h, degrades the polyester fibre which cannot be used for further processes. The treated polyester fibre was rinsed with cold water in several times to remove the residual alkali.

An experimental plan was developed by Box-Behnken method for fabrication of the composite and it was given in Table 1.

The surface morphology of the alkaline hydrolysed polyester is analysed through the Scanning electron microscope.

2.2 *Construction of Composite Membrane*

The PVP and activated carbon materials were thoroughly dispersed in water as per the experimental plan and the known weight of parallely arranged alkaline hydrolysed polyester fibre was immersed in the solution. The schematic representation of the composite preparation was shown in Fig. 1. The immersed polyester fibres are allowed for 24 h to adsorb the prepared solution and the samples are dried at 60 °C for 5 h to remove the residual water from the constructed composite. The dried sample was tightened across its axis to obtain the required dimension of the composite. This

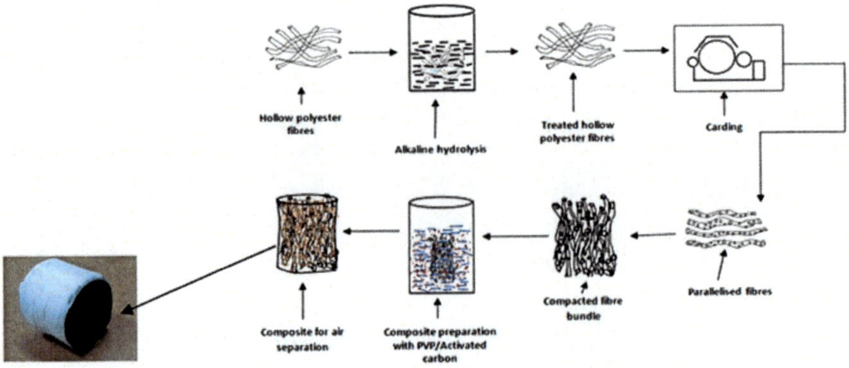

Fig. 1 Schematic diagram of composite preparation for air separation

process was followed for all samples to obtain uniform volume of the composite as 100 cm^3.

2.3 *Construction of Air Separation Unit*

The construction of air separation unit comprises pressure gauge, flow meter with regulator, non-return valve and other components. The flow meter with regulators is used to control the feed air to the separation system. The output air was collected in two separate valves (Permeate in one valve and Retentate in another valve). The schematic diagram of the air separation unit is shown in Fig. 2.

Non-return valve (NRV) is placed in the feed side that normally allows the air to flow through it in only one direction. The compressed air with the pressure of 5 psi is passed through the pre filter and also passed through the membrane with the rate of 5 lpm and the permeate was collected separately.

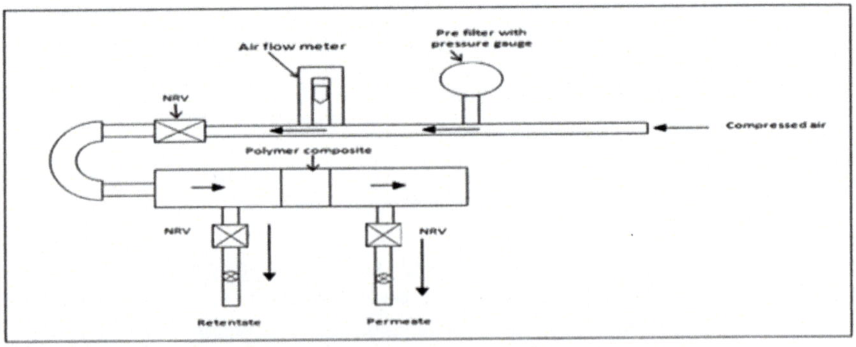

Fig. 2 Schematic representation of air separation unit

2.4 Gas Permeation Theory

Gas separation in membrane occurs due to different levels of solubility and diffusion of the gaseous components. The mass transport of the gas through the membrane is given by Fick's law

$$N = P_r \frac{A}{\delta} \Delta p \tag{1}$$

where,

N flow rate of gas, cm^3 (STP)/s.
P_r rate of gas permeability.
A membrane area, cm^2.
δ membrane separating barrier thickness, cm.
ΔP trans partial pressure difference of gas, cm–Hg.

The rate of separation of particular component between gases will depend upon the relative permeabilities of the gases to be separated. This is expressed as alpha (α) and

$$\alpha = \frac{\text{permeability of } O_2}{\text{permeability of } N_2} \tag{2}$$

ORSAT apparatus was used to evaluate the purity of oxygen in the permeant. It is a laboratory gas analysing equipment which is used to analyse gas samples like oxygen, carbon monoxide and carbon dioxide content.

3 Results and Discussion

3.1 Effect of Alkaline Hydrolysis on Fibre Surface

The alkaline hydrolysis of polyester has been carried out by varying the concentration of NaOH and treatment time. The surface change of fibres has been observed by scanning electron microscope and the images are given in Fig. 3a–c. The Figure shows that the modified surface of the polyester which is having pores and cracks on the surface. The intensity of the pores is increased by increasing the treatment time.

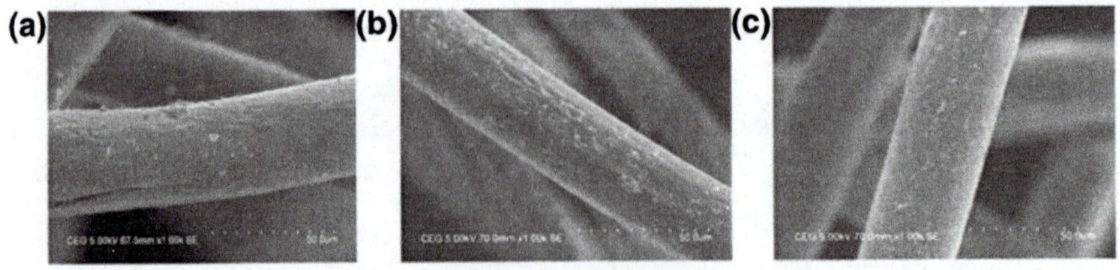

Fig. 3 SEM images of surface treated hollow polyester with NaOH for **a** 1 h, **b** 2 h and **c** 3 h

3.2 Weight Loss Due to Alkaline Hydrolysis

The degradation of polyester is the result of the alkaline hydrolysis which is influenced by the concentration of NaOH, treatment time and treatment temperature. In this process, the optimised treatment temperature as 60°°C and the concentration as 25% were used to treat the samples. In the higher treatment time, there will be reduction of the fibre length as 3–5 mm due to degradation. (at 4 h and 6 h shows much higher degradation). Upto the treatment time of 3 h, no breakage of fibres was observed. Hence, the treatment time has been taken from 1 to 3 h to hydrolyze the samples which can create pores/cracks on the surface of hollow polyester fibres. Due to hydrolysis, it is observed that higher weight loss in fibre at higher concentrations of NaOH.

3.3 Characterisation of Activated Carbon

Activated carbon was used in this process to blend with Polyvinylpyrrolidone to enhance the separation of oxygen. The particle size of the activated carbon was analysed using laser scattering particle size distribution analyser which yielded a mean size of 26 micron.

3.4 Air Separation Efficiency

The separated oxygen from the atmosphere air was tested with ORSAT apparatus and permeability rate of air through the membrane was calculated. The results are influenced by Alkaline hydrolysis time, percentage of activated carbon and PVP and it is given in Table 2.

It was observed that the higher O_2 percentage in the permeate was achieved as 38.62%. This was due to higher permeability of O_2 and selectivity of O_2/N_2 by the combination of PVP/Carbon. The rate of separation was also achieved as 4.1 l/min due to highly porous nature of the activated carbon and also more pores on the surface of the treated polyester. The PVP as 10% and activated carbon as 6% gives

Table 2 Oxygen separation and air separation rate

S. No.	Treatment time h (X_1)	Conc. of PVP % (X_2)	Activated carbon % (X_3)	Oxygen (%)	Air separation rate l/min
1	1	5	4	26.57	3.62
2	3	5	4	33.21	3.90
3	1	15	4	35.22	3.76
4	3	15	4	38.62	4.10
5	1	10	2	29.66	3.74
6	3	10	2	35.46	3.96
7	1	10	6	33.22	3.85
8	3	10	6	37.07	4.22
9	2	5	2	27.96	3.70
10	2	15	2	36.3	3.96
11	2	5	6	30.89	3.90
12	2	15	6	31.98	3.92
13	2	10	4	29.35	3.83
14	2	10	4	29.5	3.58
15	2	10	4	29.35	3.56

37.07% of O_2 in the permeate and the rate of air separation was given as 4.22 which is maximum due to higher treatment time and higher concentration of activated carbon. It was observed that higher the PVP content gives higher permeability of O_2 and higher the treatment time and activated carbon percentage results higher the flow rate of air through the composite. In this process the α value varies from 0.36 to 0.62.

3.5 *Effect of Variables on Oxygen Enrichment and Air Separation Rate*

The increase in concentration of PVP and treatment time increases the oxygen percentage in permeant was shown in Fig. 4a. Both independently influences the enrichment of the oxygen in the permeant. Fig. 4b, c show that the contribution of activated carbon is minimum for enriching the oxygen in the permeant while comparing the PVP. Hence, the impact on oxygen enrichment of the variables is followed as PVP > treatment time > activated carbon.

The rate of separation of oxygen through the membrane was attributed by the higher O_2 selectivity nature of the PVP. The air flow rate of the membrane was attributed by the percenage of activated carbon. The rate of air flowthrough the membrane was influenced by the amount of activated carbon and PVP loaded. Figure 5a–c show that the influence of treatment time on air flow through the

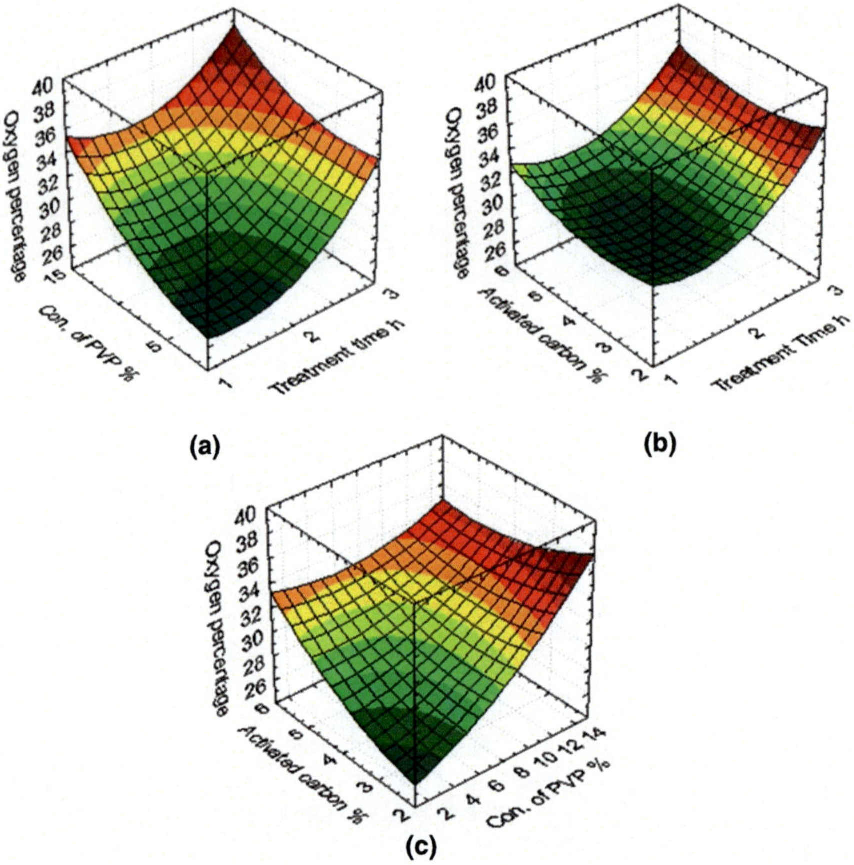

Fig. 4 **a** Effect of PVP% and time of treatment on oxygen enrichment % and **b** Effect of activated carbon % and time of treatment (h) on oxygen enrichment % and **c** Effect of PVP% and activated carbon % on oxygen enrichment%

membrane was more due to the formation of more pores on the surface of the fibres. The lower concetration of PVP and higher concentration of activated carbon also leads to increase in air flow through the membrane due to the porous nature of the activated PVP.

4 Conclusion

Separation of oxygen is an important process for commercial application and the performance of oxygen separation membrane varies depending on the process variables. In this work, a new concept has been used to construct the air separation

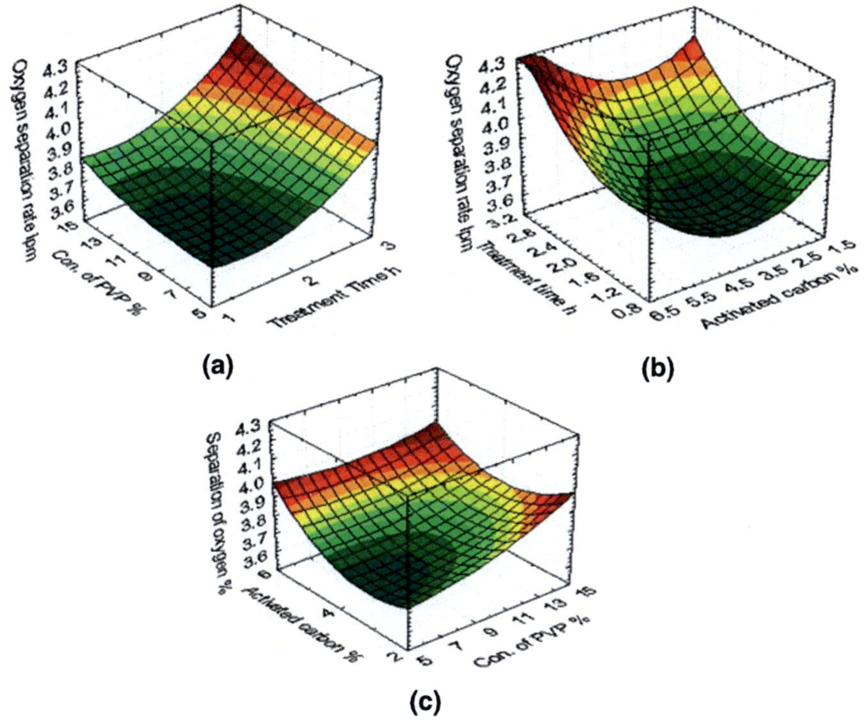

Fig. 5 **a** Effect of time of treatment and Conc. of PVP on oxygen separation rate and **b** Effect of time of treatment and Conc. of activated carbon on oxygen separation rate and **c** Effect of Conc. of PVP and Conc. of activated carbon on oxygen separation rate

membrane. Due to the various advantages of hollow fibres, in this work PVP and activated carbon impregnated hollow polyester fibre membrane has been constructed. To achieve the pores on the surface of the hollow polyester fibre, alkali treatment was given to the fibres and the process was optimised. Various proportions of PVP and activated carbon were incorporated as per the experimental plan. The constructed membranes have been tested for its functional and physical properties. The constructed membrane has shown higher air separation rate and oxygen purity. The time of treatment influences the weight loss on polyester and improves the flow of oxygen through the pores present in it and the concentration of PVP content influence the oxygen purity. Process variables of 3 h NaOH treated sample with 15% PVP and 4% activated carbon separates air with higher oxygen purity. 3 h NaOH treated sample with 15% PVP and 6% activated carbon has high rate of air separation. Hence the new concept of construction of membrane has given oxy rich air for the commercial applications.

References

1. Fabian C, Bimbi L, Pizzichini M, Santarossa L (1996) Gas Sep Purif 10:75
2. Fernandez-Barquin A, Casado-Coterillo C (2016) Susana valencia and angel Irabien. J Membr Sci 6:1
3. Schmidt WP, Winegardner KS, Dennehy M, Castle-Smith H (2001) Pro Safety Prog 20:269
4. Aryegollan, Kleper HM (1984) J Ind Energy Technol 84:298
5. Yan L, Yu Y, Li Y, Zhang Z (2010) J Sch Mech Eng 9:2453
6. Burdyny T, Struchtrup H (2010) J Energy 35:1884
7. Sridhar S, Bee S, Bhargava KS (2001) J Chem Ind 1
8. Yampolskii PY, Belyaev AA, Drioli E (2005) J Chem Eng 1:1
9. Wang X-Y, Hill JA, Freeman DB, Sanchez CI (2008) J Mem: Sci 314:15
10. Coker TD, Prabhakar R, Freeman DB (2003) J Chem Engg Edu 1:60
11. Yang W, Wang H, Zhu X, Lin L (2005) J Sci 35:155
12. Javaid A (2005) J Chem Engg 112:219
13. Tremblay P, Savard MM, Vermette J (2006) J Membr Sci 282:245
14. Ismail AF, Norida R, Sunarti AR (2002) J Sci Technol 24:1025
15. Souza VC, Quadri MGN (2013) Brazilian J Chem Eng 30:683
16. Bastani D, Esmaeili N, Asabollahi M (2013) J Ind Eng Chem 19:375
17. Sanders FD, Smith PZ, Guo R (2013) J Polymer 54:4729
18. Orme CJ, Klaehn JR, Stewart FF (2004) J Membr Sci 238:47
19. Phair JW, Badwal SPS (2006) Sci Technol Adv Mat 7:792
20. Chong KC, Lai SO, Thaim HS, Teoh HC (2016) J Eng Sci Technol 11:1016

A Novel Approach Towards Design and Development of Indian Men's Workplace Casual Footwear with Specific Reference to Sizing

Check for updates

Sivasakthi Ekambaram and Chitra Arora

1 Introduction

India's annual footwear consumption stands at 2.1 billion pairs and it is the third largest footwear consumer, globally after China and the USA and has recorded a healthy growth over the last decade, driven by the rise in income levels, growing fashion consciousness and increasing discretionary spending [1]. The domestic footwear industry in India is valued at approximately ₹. 22,697 crores (US $ 319.27 crores), projected growth at 11–12% per annum. Footwear retailing constitute about 9% in the total consumer market [2]. The Indian footwear market can be bifurcated into men, women and kids' segments. Men's footwear occupies the maximum share in the Indian footwear market. The share of men's footwear was around 54% in 2016, followed by women's footwear which was around 37%. Further, kids' footwear was around 0.9% in the market [1]. The footwear industry in India is further bifurcated into casual footwear, mass footwear, premium and sports footwear. Mass footwear usually refers to low price footwear and majorly consists of slippers. On the other hand, casual footwear involves those preferred by people for daily wear in schools, colleges or workplace, etc. Casual footwear dominates the market followed by mass footwear. The share of casual footwear was estimated to be 61% in 2012. However, casual footwear is expected to continue to dominate the market, the share of sports and premium footwear is expected to increase [3]. The common styles of men's workplace casual footwear are moccasin, boot, derby, brogue and sneaker [4].

S. Ekambaram (✉)
National Institute of Fashion Technology, New Delhi, India
e-mail: sivasakthi.ekambaram@nift.ac.in

Mewar University, Chittorgarh, India

C. Arora
Institute of Home Economics, University of Delhi, New Delhi, India

V. Midha and A. Mukhopadhyay (eds.), *Recent Trends in Traditional and Technical Textiles*, https://doi.org/10.1007/978-981-15-9995-8_10

The study by Anand and Alekya [3] analyses the consumption pattern of footwear consumers restricted to the twin cities of Hyderabad and Secunderabad. People consider comfort as the primary important factor while purchasing footwear. While men consider quality as the second important factor. Young people prefer to have multiple footwear of different variety. Other age groups prefer to have one pair but which is comfortable and long lasting. The case study by Sarkar [5] discusses the challenges faced by the domestic footwear sector. The study by Aout et al. [6] advocates barefoot walking, which helps to preserve natural foot function. When the substrate does not allow for barefoot locomotion, footwear should be worn that protects the foot from injury, but that is unrestrictive enabling the foot to function as much as possible as in the unshod condition. The review by Rossi and Tenant [7] discusses the shoe fitting procedure comprehensively along with outlining fifteen test points for footwear fit. Delhi Economic Survey [8] lists the workplaces with reference to the National Capital Territory (NCT) of Delhi. The research by Chien-Chung [9] discusses in detail three existing foot measurement systems in the UK footwear industry and also the differences among these three measurement systems. The Shoe and Allied Trade Research Association (SATRA), Kettering, UK the biggest footwear technology research centre in the world, offers a foot measuring system for use in footwear research and development. Tomassoni et al. [10] outlined the inclusion criteria for their study which analysed the main morphological parameters of foot of a large population of individuals of different ages, to provide basic information for the development of appropriate footwear for the elderly. The research report by Clapham, SATRA [11] offers a surer means of measuring dimensions of the feet for arriving at correct last dimensions. ISO 8559:1989(en) [12] includes the procedure for taking girth measurements of the leg which are essential for the manufacture of boots, which is one of the styles of Indian men's workplace casual footwear. IS:1638–1969, [13], Bureau of Indian Standards—this standard prescribes the sizes and fittings for footwear last which are required for manufacture of footwear along with the specifications with respect to foot and last dimensions and;

1. The approximate equivalent shoe sizes in Paris Point size scale, English size scale and American size scale.
2. Formulae for interconversion of Paris Point size and English size.

4.0 Industry Standards promotes the use of smart technology and product design based on user centred parameters, etc. AK A64—WMS, Germany provides shoe last design guidelines. The developed shoe last has three available widths viz.

- W (Weit = Wide).
- M (Mittel = Medium).
- S (Schmal = Narrow).

The shoe system is quite elaborate on the foot dimensions. The dimensions are measured in millimetres. It also outlines the guidelines to be followed by last designers to develop the flattened last bottom profile/plan. A set of prescribed points on the feather edge/line could be moved by the minimum amount, so that the resulting last bottom surface would flatten to best fit [14].

Footwear sizes are one of the most complicated and erratic areas of the whole footwear fitting process. And no matter the experience of the fitter, footwear sizes are one of the least understood, or the most misunderstood, elements in footwear fitting. While size is obviously important, we tend to overmagnify its importance by the assumption that if the footwear is the right size it will automatically deliver the proper fit. This of course is quite untrue. From the stand point of the fitter, footwear size consists primarily of two measurements: overall length and ball width. These measurements however do not indicate the true size of the foot or the footwear. Heel width, waist girth, instep, heel to ball, arch length also come in "sizes". The footwear is presumably "sized" in its various sections to match the corresponding sections of the foot. At least that's the way it is supposed to be. But often it doesn't happen that way [15].

Due to the lack of Indian sizing surveys, footwear manufacturers are forced to use the size charts of other countries where the anthropometric make of the population is very different from that of India. This results in manufacturing of a product which might not fit the Indian feet types.

Few systems abroad, also take the width/girth of the feet into account. Some regions use different footwear size systems for different types of footwear—men, women, children, sports or safety footwear. Each size of footwear is considered suitable for a small interval of foot lengths. In order to accommodate variations in foot sizes, alternative width fittings are available. By the use of fittings it is possible to fit greater number of feet more efficiently. The number of fittings used will be governed by the class of trade being catered for [16].

Only a few retail brands like Bata, Action, Liberty, Relaxo, J. D. Williams offer a range of fittings for each size in India. NIKE seems to offer customized trainers and sneakers in India, in the price range of ₹ 8000–27,000 (US $ 120–406). When footwear of different width options is not available for the same length then people with wider feet are forced to purchase shoes of larger sizes while those having narrower feet will have to wear too wide shoes. In both cases feet may suffer, but certainly shoes will go out of their shape much faster than in case of normal fit and their durability will also be reduced.

The comfort of shoes seems to affect the fatigue, the injury outcome and the performance of each person [17]. One of the major causes of foot debilitation is the wearing of ill fitting and improper footwear which causes blisters, corns and other foot injuries which if left untreated could lead to limb impairment and damage. To avoid such an occurrence, it becomes imperative to design and develop proper and correctly fitting footwear and the major ingredient in the success and provision of correctly fitting footwear is the "last", which is a three dimensional form of the foot and on which the footwear would be made [18]. The shoe last is a model representation of the feet—a kind of average of the given person's or population's feet. This lends the correct shape, fitting and dimensions to the footwear. Information on the shape and dimensions of the foot is of key importance in ensuring that footwear is designed to fit correctly. As with all physical objects, the human foot can be characterized by its dimensions. The appropriate set of measurements, processed according to established techniques is used in the development of sizing systems

[19]. In the footwear industry it is essential to have statistical data of the proportions of the foot of the local population. This is essential for last development as lasts designed and manufactured in other countries cannot serve their purpose in India, owing to differences in population, climate, wearing conditions, urbanization, etc. Last manufacturers use at least 30 dimensions to build a foot last [15]. The dimensions of the lasts in present use were valid for a population representative of the United Kingdom or France decades ago. Fit and comfort, i.e. consumers' satisfaction as well as design and pattern engineering, tooling of the production, component supply (coordination) all depend on the successful design of the shoe last. For this purpose, accurate data are required on the geometry of foot. At the present time, characterized by a transition to a free market economy, this requires even more urgency because the products have to stand up to strong competition from all over the world.

Thus, the main aim of the present study is to come up with a strategy for providing well-fitting footwear to Indian office going men. In order to achieve this identified anthropometric dimensions of feet of Indian men, were collected. It is expected that the survey would.

- Reveal differences among various ethnic groups—if they exist or prove that no differentiation is needed when footwear is produced and supplied.
- Set basic measurements to be used for marking sizes of footwear, establish size ranges providing the required coverage of the population with footwear to be produced using industrial technology for retail.
- Produce rules and numerical databases for designing well-fitting shoe lasts which will avoid development of static and other foot diseases caused by wearing shoes which are not comfortable.

The results of anthropometric measuring provide an insight into the occurrence and regional distribution of certain sizes in the investigated population. The information and results of the foot measurement system can also be used in research, made to measure services, retail studies, demography, ergonomy, etc. [19].

2 Experimental

Due to the lack of up to date published material related to the subject of research it has been necessary to visit the shoe and last manufacturers in order to gain a better understanding of the production processes and standard model last design and making techniques. Consultations and interviews were conducted with specialists and experts from the factories and experienced faculty members in footwear development. Some related research centres and manufacturers were also visited. A series of informal visits to experienced engineers and technicians, and discussions with supervisors were first carried out. These resulted in the modification of the area of enquiry and the identification of additional informants or sources of information.

Experts in the field were consulted through a semi-structured interview schedule to.

1. Validate the existence of the problem of footwear sizing for Indian men's workplace casual footwear in the NCT/NCR region.
2. Understand the methodology of fit and field trials.
3. Ascertain availability of infrastructure, facilities and other resources with respect to anthropometric survey of feet in NCT/NCR, last assessment and design in NCT/NCR regions, development of customized shoes in NCT/NCR regions and footwear comfort testing, insole pressure measurement, etc.

Further primary and secondary data has been collected for a pilot study to ascertain problems related to men's footwear, in Indian retail market. The total sample size of 71 respondents was drawn from a variety of consumer groups who have diverse demographic backgrounds and societal lifestyles. The area of the study was primarily National Capital Region (NCR).

Further identified anthropometric dimensions of feet of Indian men, have been collected following SATRA foot measurement system for both the right and left feet. Since styles of workplace casual footwear includes boots, certain anthropometric dimensions of both the legs were also collected using the procedure outlined in ISO 8559:1989(en) standards. Morphological characteristics included thirty-one anthropometric dimensions of feet and leg of Indian men, which in turn included four angle measurements, ten length measurements, ten girth measurements, four height measurements and three width measurements [9, 12].

After thorough review of existing literature in the area, the list of anthropometric measurements to be collected was identified and the required list of instruments/tools, stationery, consumables were also identified and mobilized. A foot anthropometric measurements/dimensions chart was designed and drafted along with a voluntary consent form which each subject would sign.

Subjects. New Delhi, the capital of India, sprawled over the west bank of the river Yamuna is one of the fastest growing cities in India. It is surrounded on three sides by Haryana and to the east, across the river Yamuna by Uttar Pradesh. As per 2011 census the population of National Capital Territory (NCT) is 1,67,53,235 out of which the male population is 89,78,154 and female population is 77,75,081.

Migrants from other states constitute a sizeable portion of Delhi's population. Employment factor and the capital's strong labour market is the major reason for migration of the population, especially men, from other states to Delhi [20]. The migrants are mainly from the states of Uttar Pradesh (UP), Haryana, Bihar, Rajasthan, Punjab, West Bengal (WB), Madhya Pradesh (MP) and other states [21]. The user is either of middle or top management level for whom there is usually no specified uniform, even though a dress code might exist.

Inclusion criteria. All participants were to meet the following inclusion criteria:

(1) No history of congenital deformity in the lower extremity or foot;
(2) No previous history of lower extremity or foot fractures;
(3) No surgical operation on foot and lower extremity;
(4) No systemic diseases that could affect lower extremity or foot posture;

(5) No history of trauma or pain to either foot, lower extremity or lumbosacral (lower back) region at least 12 months prior to start of the investigation [10].

The survey was also restricted to people whose differences between the left and right feet were not visible in the first sight, the person can walk and run without any difficulties [19].

2.1 Measurement Methodology

Anthropometric foot measurement instrumentation. The following is the list of tools, instruments, stationery etc., used for measuring and recording the foot data (Figs. 1, 2, 3 and 4).

Fig. 1 Stature measurement scale

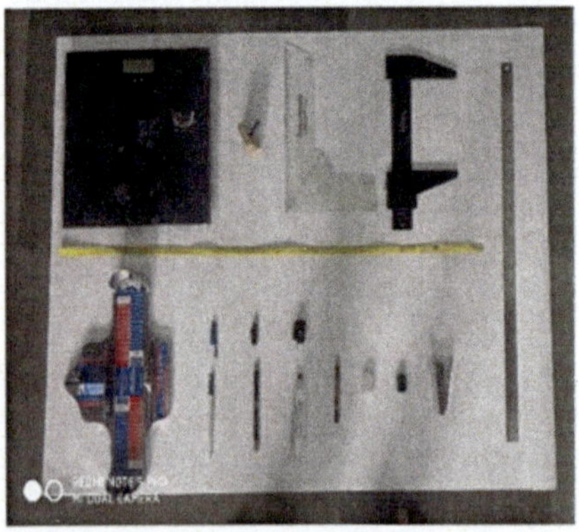

Fig. 2 Weighing scale, in-house developed scriber block, shoe makers square, small bone anthropometer, footwear measuring tape, combination measuring device, pen, eye liner pencil, marker, HB pencil, eraser, sharpener, stapler and steel scale

Fig. 3 Graph sheet

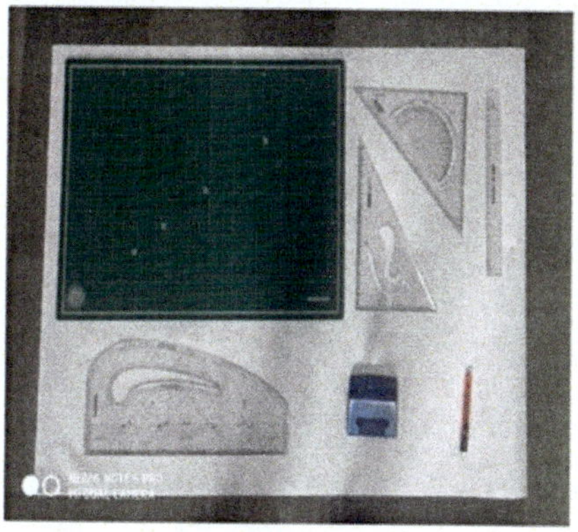

Fig. 4 Cutting mat, protractor, set square, scale, pattern master, double punch and paper cutter

1	Properly designed chart
2	Thermocare height measurement scale for baby and adult 2 m
3	Body weight weighing scale
4	Scriber block
5	Shoemakers square © 1995
6	Footwear measuring tape
7	Small bone anthropometer,
8	Combination measuring device
9	Ball point pen, marker and eye liner pencil
10	Cutting mat

(continued)

(continued)

11	Graph sheets A3 size,
12	HB Pencil, eraser, sharpener, stapler
13	Steel scale, set squares, protractor

The instruments and tools. Thermocare height measurement scale for baby and adult 2 m was mounted on the wall, which was covered with white garment pattern making sheets. The other instruments and tools like body weight weighing scale, in house developed wooden scriber block, shoemakers square © 1995, footwear measuring tape, small bone anthropometer, combination measuring device etc. are portable, so data collection relating to determining shoe size ranges, feet variation by geographic regions, etc. were carried out on the spot with minimum efforts wasted in setting up the system.

The measurement methodology consisted of a precision manual measuring procedure where the subjects' feet and leg are measured as per the procedure outlined in SATRA foot measuring system and ISO 8559:1989(en) standards.

Descriptive statistics/demographic and personal characteristics such as:

1	Date of birth and age
2	Stature/height
3	Body mass
4	Occupation/employment details
5	Ethnic group

were recorded through questionnaire and by using wall mounted height measuring scale and body weight weighing scale.

Morphological characteristics/foot anthropometric measurements/dimensions. The feet and leg of the subjects were felt for some bony and anatomical points. Six anatomical points on the foot and four anatomical points on the leg were marked with an eye liner pencil. The foot plan of both the feet, were drawn on suitable size graph sheet using the in house made wooden scriber block and HB pencil. The corresponding positions of the six anatomical points on the foot were marked on the foot plan using steel scale and shoe makers square. Morphological characteristics which include thirty-one anthropometric dimensions of feet of Indian men, which in turn included four angle measurements, ten length measurements, ten girth measurements, four height measurements and three width measurements were measured using protractor, scale, measuring tape and small bone anthropometer. The size heel to toe and heel to ball and fitting of the foot was measured using combination measuring Brannock device.

Measuring conditions. The most important condition is the body position of the subject while the measurements were taken as well as when the girths were measured. The person stood on both feet, whereby the two feet were placed parallel approximately in a distant equivalent to the shoulder span (width), while the body weight was equally distributed between the two feet. Both the feet of the subjects were measured in this survey.

The measuring procedure. After recording the descriptive statistics/demographic and personal characteristics in the properly designed foot anthropometric measurements/dimensions chart, the information on confirmation for the inclusion criteria were recorded. The height and weight were measured using the appropriate instruments. The anatomical points on both the feet and leg which define the foot measures were marked using the eye liner pencil. The foot plan of the feet were drawn using the in-house developed wooden scriber block and pencil on appropriate size graph sheet and the positions of the anatomical points were also marked on the foot plan. The girths were measured using a footwear measuring tape. Angle measurements were measured from the foot plan using protractor. Length and width measurements were measured using scale. Certain height measurements were measured using a small bone anthropometer. The size and fittings were noted using combination measuring device. The longest toe, i.e. the big toe or the second toe, etc. was noted on the foot anthropometric measurements/dimensions chart. After ensuring that all the required measurements were recorded, the foot anthropometric measurements/dimensions chart of individual subjects, the graph sheets on which the foot plan is drawn and the consent forms were filed in the dedicated box files. Measurements of different morphometric parameters for each subject were recorded in millimetres and the values obtained from the right and left foot were summed up and divided by two. Hence values on which statistical analysis are to be conducted refer to the mean value of right and left foot parameters under evaluation.

3 Results and Discussion

3.1 Semi-Structured Interviews of Experts in the Field Were Conducted and the Following Are the Outcome of the Research Conducted so Far

According to Debasish Das Gupta (Interview, August 28, 2017) faced with problems of non-standard size grading etc., India is not able to compete with China in non-leather shoes. Due to lack of awareness the Indian consumer emphasises more on looks rather than technical aspects and sizing of footwear. In house product development of lasts is lacking in India. According to Ms. Satyam Srivasatava, Senior Consultant (Footwear Technology), Footwear Design & Development Institute (FDDI), Noida (Interview, August 30, 2017) testing facilities for SATRA Footwear Comfort Index etc., is not available in India.

According to Mr. Shravan Kumar Singh, General Manager—Product Development, Mmojah, Sonipat, Haryana (Interview, October 10, 2017), in the current scenario, emphasis is more on look rather than on sizing. Russia uses Mondopoint sizing system. Bata did a foot survey around 50–70 years back. But due to feasibility issues did not use the data and continued with the European sizing system. There is a lot of difference between the feet of the urban population in comparison to the rural one. Width increment is 1/3rd each ball girth increment. Indicative fitting is used by the manufacturer and wholesaler and real fitting is for the consumer. G-Wide and H-Very Wide fittings are mostly used in India. There are three stages for ensuring the correct fit before the bulk production starts, by the manufacturer.

1. Last.
2. Fit test—done in a lab.
3. Fit wear test—an employee of the matching size is made to wear the footwear and Quality Control inspection is done after 15, 30 and 45 days. A set of questions are asked to the employee as per a prescribed format.

There is an average of 15 mm gap at the toe portion between the feet and the shoe. Only 1% complaints received from footwear consumers during after sales service are genuine in India.

Mr. Kripal Singh, Top Lasts, Noida (Interview, December 07, 2017) explained the principles of last making.

Each last manufacturing company follows its own single colour for last, because changing colour is difficult. Top lasts uses RECAD—a Poland based software.

3.2 The Primary Data for the Study Was also Collected from a Random Sample Using a Questionnaire. The Respondents of Various Age Groups, Income Levels, Occupations Were Asked for Their Consumption Pattern

(See Table 1).

- 42% respondents reported to have a foot injury/illness. The occurrence of foot injuries in descending order is heel pain followed by heel spur, Achilles tendonitis, arch pain, bunion, callus, claw toe and Morton's neuroma.
- The respondents' preference of style of men's workplace casual footwear in descending order is Derby followed by boot, sneakers, brogue and moccasin. On an average each respondent possessed 5 pairs of footwear at any given point of time. Maximum number of respondents used leather footwear. The respondents' preference of brands for men's workplace casual footwear in descending order is Woodlands followed by Bata, Lacoste, Liberty, Action, Lee Cooper and HRX.
- The preferred mode of purchase of footwear in descending order is through retail outlets followed by online, through factory outlets. 39% respondents availed of after sales service for their workplace casual footwear. The maximum complaints in workplace casual footwear in the descending order was on account of sole cracking/breaking followed by sizing, manufacturing defects, workmanship and colour fading of trims.

Table 1 Demographic profile

Category Age (years)	Respondents	Percentage
21–30	19	27
31–40	08	11
41–50	21	30
51–60	20	28
61–70	02	03
71–80	00	00
81–90	01	01
Monthly Income (₹)	**Respondents**	
15,000–50,000	26	
50,000–1,00,000	15	
1,00,000–1,50,000	05	
1,50,000–2,00,000	07	
2,00,000–2,50,000	04	
≻2,50,000	14	
Workplace	**Respondents**	**Percentage**
Bank	05	07
Corporate office	09	13
Educational institution	10	14
Factory	05	07
Government office	16	23
Hotel	01	01
Hospital	05	07
Media	03	04
Restaurant	01	01
Others	16	23
Ethnic group	**Respondents**	**Percentage**
Purvanchali	10	14
UP	12	17
Uttaranchal	01	01
Haryana	11	16
Bihar	08	11
Jharkhand	03	04
Rajasthan	02	03
Punjab	16	23
West Bengal	02	03
Madhya Pradesh	01	01
Other states	05	07

- The usage of sizing system in descending order is British followed by American, Europoint and French. Only 9% of respondents used width fittings. 25% of respondents faced problems with available sizes/sizing in workplace casual footwear.

3.3 Anthropometric Survey

Statistical analysis. The task is to select those measurements which can determine all others, whereas the number of basic measurements should be kept to the minimum. If there is a strong relationship between one of the measured leg or foot dimensions and all others then the given property can be treated as the major size of the foot or shoe. The history of shoemaking had selected the foot length, the ball girth and/or width as such basic data out of which all other parameters required for designing shoe lasts, shoe upper and bottom patterns can be derived either by using (simple) equations or appropriately constructed tables. A large number of anthropometric surveys made in several countries also proved that measurements oriented in the same direction of the space (axes of the Cartesian system of coordinates) correlate very well (e.g., instep or waist or heel measure have strong correlation with the ball girth). The foot and leg survey should either verify the existing practice or find better, stronger, simpler and/or more reliable rules. The ideal correlation and regression analysis would investigate all possible combinations of the data fields (measurements and personal data) and various types of relationships (e.g.,) linear, quadratic or higher degree of paired, multivariable linear equations). There are three major directions in foot: length, width, while relationship of the measurements in the third direction (height) with other needs to be ascertained. Correlation coefficients for the same are to be calculated and correlation tables created. Histograms belonging to frequency variable in the new tables to be generated. Using the statistical Means of these tables we could select the so-called middle size values and calculate the typical geometric data for last modelling.

Size table to be created with frequencies, mean and deviations for English sizing system. The average foot length, the standard deviation, the size increment and the batch volume are to be computed. A batch may be a complete order or its fraction, the number of pairs in the production lot, the amount of footwear kept together in handling, packaging, shipping, transporting, storing, ordering, etc. The footwear industry uses the typical batch sizes which are 6, 10, 12, 60, 100, 120, 144, 500, 600, 1000 pairs. The method to be used for testing differences between samples from different geographic and/or ethnic regions is based on Student's t distribution. To increase the reliability of retail sale, i.e., maximize the likelihood of serving every consumer with well-fitting footwear with minimum size inventory, then the distribution of characteristic anthropometric measurements in regional sub groups should also be compared. This would be done using the F test. Data are first grouped by foot length expressed in a given size system (in our case in English half sizes), then average ball girths and widths are computed for each size. Mean and deviation of stature, weight, stick length, plan length, ball width, seat width etc. for the given size number are to be computed. Now the increments in girth/width can be computed for each adjacent pair of length sizes: the weighted average of these Figs will give the computed (theoretical) increment by (English half) sizes.

Discussion

The key challenge to the referred study was to define and identify statistically relevant sample size and further ensure that the data collection is from the relevant sample. It would, moreover, not be feasible as the majority of the paired comparisons are meaningless in footwear technology. Complicated rules—even with appropriate computer programmes—would hardly be used by shoe designers and marketing department. Domestic vendors especially small players, order footwear in small batches and at less regular frequency. In this scenario, the size of market, type and quality of product to be manufactured broadly determines the choice of production technology in different market segments. Due to the cost involved in investing in a specific sizing system for the market under reference, industry may not see it as a commercially viable option. Redesigned footwear to fall within the financial affordability of the consumer. Availability of resource with regard to state of art laboratory facilities within the vicinity of geographical region where research was conducted to assess long-term comfort of the shoes constructed on the redesigned lasts was also an issue.

4 Conclusion

Capacity building in product development of footwear lasts would in a great way lead to use of more comfortable footwear by Indian office going men. India's dependence on foreign players for quality footwear would also be reduced. Though there are lot of factors and costs involved in following a better and foolproof footwear sizing system, a beginning has to be made which would lead to lot of technological advancements in the area.

Acknowledgements The authors express their sincere gratitude to all the subjects. We would like to thank Mr. Praveen Kumar of National Institute of Fashion Technology, Delhi campus for developing the wooden scriber block used for drawing the foot plan of the subjects on the graph sheets. The authors also express their sincere gratitude to Ms. Harvinder Kaur of National Institute of Fashion Technology, Delhi campus for the support rendered towards photography while taking the anthropometric measurements.

Competing Interests The authors declare that they have no competing interests.

References

1. HDFC Bank Investment Advisory Group, Footwear sector. Sector update. (2018). https://www.hdfcbank.com › Sector-Update-Footwear-Sector-June-2018 Accessed 08 Oct 2019
2. Reliance. About Indian footwear industry. (2019). indiatradefair.com › uploads › About_Indian_Footwear_Industry_14_05_... Accessed 08 Oct 2019
3. Anand, G., Alekya, U.: A Study on Consumer Behaviour in the Footwear Industry. International journal of applied services marketing perspectives, vol. 4, pp. 1768–1775. Retrieved from https://pezzottaitejournals.net/pezzottaite/images/ISSUES/V4N3/IJASMPV4N313.pdf. (2015)
4. https://www.menshealth.com/style/5-shoes-every-guy-should-own/slide/6
5. Sarkar, S.: A Case Study of Footwear Industry in India. Retrieved from https://www.ihdindia.org/Formal-and-Informal-Employment/Paper-4-A-Case-Study-of-Footwear-Industry-in-India.pdf. (2011)
6. Aout KD, Pataky TC, De Clercq D, Aerts P (2009) The effects of habitual footwear use: foot shape and function in native barefoot walkers. Footwear Sci. 1(2):81–94
7. Rossi WA, Tennant R (2013) Professional Shoe Fitting, National Shoe Retailers Association, ISBN 9781463647384. Tucson, Arizona
8. https://www.indiaenvironmentportal.org.in/files/file/ESEngcompressed.pdf
9. Chien-Chung, C.R.: An Investigation into Shoe Last Design in Relation to Foot Measurement and Shoe Fitting for Orthopaedic Footwear. (A thesis submitted for the degree of Doctor of Philosophy in the department of Medical Engineering and Physics). King's College School of Medicine and Dentistry, University of London, London, United Kingdom. (1993)
10. Tomassoni, D., Traini, E., Amenta, F.: Gender and age related differences in foot morphology. Maturitas (2014). https://doi.org/10.1016/j.maturitas.2014.07.019
11. Clapham, J.C.R.: Children's and Adolescent's Foot Measurements. Abridged version of Research report No. R R 114. The British Boot, Shoe and Allied Trades Research Association (1957)
12. International Organization for Standardization https://www.iso.org/obp/ui/#iso:std:iso:8559:ed-1:v1:en
13. Bureau of Indian Standards.: Specification for Sizes and Fitting of Footwear, IS:1638–1969. (1970)
14. Luximon A, Luximon Y (2017) She-last design and development. The Science of Footwear. CRC Press, Boca Raton, London, New York, pp 203–204
15. Rossi W (1988) The futile search for the perfect shoe fit. J. Test. Eval. 16(4):393–403. https://doi.org/10.1520/JTE11084J
16. Salminen, T.N.: Footwear Sizes and Fittings. Indian Institute of Leather Products: Chennai. P. 2–4, 7–11, 18–9 (1992)
17. Kalebota, N., Szirovicza, L., Tucakovic-Mujagic, L., Mirko Drenovac, M.: The Structure of Body Measurements for the Determination of Shoe Sizing for Young Croatian men. Collegium Antropologicum. 27:635–43. UDC 572.512-055.1:685(497.5) 2:635–643 (2004)
18. Luximon A, Luximon Y (2009) Shoe-last design innovation for better shoe fitting. Comput. Ind. 2009(60):621–628. https://doi.org/10.1016/j.compind.2009.05.015
19. Gopalakrishna G (2014) Anthropometric Characterization of Indian Feet (Doctoral thesis). Anna University, Chennai
20. Dupont V (2000) Spatial and demographic growth of Delhi since 1947 and the main migration flows. In: Dupont V, Tarlo E, Vidal D (eds) Delhi Urban Space and Human Destinies. Manohar Publishers and Distributors, New Delhi, pp 229–239
21. Department of Urban Development.: Government of Delhi, IL&FS Ecosmart Limited,City demographic profile. JnNURM. https://ccs.in › sites › default › files › files › Ch02_City Demographic Profile Accessed 19 Oct. 2019

The Effect of Wet Processing on the Comfort and Mechanical Properties of Fabrics Made from Cotton Fibres and Its Blends with Modal and Tencel Fibres in Weft

Devanand Uttam and Garima Ahlawat

1 Introduction

To survive in throat-cutting competitive environment and to satisfy modern customers; the textile manufacturers and researchers are searching some new products which have improved properties. Modern customers consider comfort as one of the most important attributes in their purchase of textile and apparel products [1]. To achieve improved functional textiles; one attempt is to produce cotton/modal blended fabric and cotton/tencel blended fabrics.

Cotton is a natural cellulosic fibre while Modal and Tencel Lyocell are regenerated cellulosic fibres. The Modal® and Tencel® LF are cellulosic regenerated fibres manufactured by Lenzing Company, Austria. Like viscose, Tencel is produced from wood pulp but the process is chemically much less complex. Modal is a type of high wet modulus rayon and manufactured using cellulose obtained from beech trees [2–4]. The Tencel fibre has highest strength, elongation and moisture regain followed by modal fibre and cotton fibres. Although, wet tenacity reduced in both Modal and Tencel fibres; while it increases in cotton fibre. The cotton fibres show highest wet tenacity followed by Tencel fibres and modal fibres [2, 4]. The cross-sectional shapes of various fibres are shown in Fig. 1.

Mostly all cellulosic fabrics are gone through the two main wet processing stages: scouring and dyeing. These may alter the structure and properties of fabrics. This requires a detailed study. In the present work, three types of plain-woven fabrics are

---

D. Uttam (✉)
Department of Textile Engineering, Giani Zail Singh Campus College of Engineering and Technology, Maharaja Ranjit Singh Punjab Technical University, Bathinda, Punjab 151001, India
e-mail: d_a_uttam@yahoo.co.in

G. Ahlawat
Department of Fashion Technology, Government Polytechnic, Hisar, Haryana 125001, India
e-mail: garima.ahlawat@rediffmail.com

V. Midha and A. Mukhopadhyay (eds.), *Recent Trends in Traditional and Technical Textiles*, https://doi.org/10.1007/978-981-15-9995-8_11

Fig. 1 Schematic diagram showing the cross-sectional view of different fibres

made using cotton yarn, cotton/tencel and cotton/modal blended yarns in weft only. The effect of wet processing on the comfort properties of these fabric samples is studied in three stages, i.e. greige state, after scouring and after dyeing. The fabric physical factors (cover factor, fabric thickness, fabric weight), air permeability, moisture transport properties, fabric stiffness and tensile properties of the fabrics were evaluated. The results are statistically analysed at 95% confidential limits.

2 Materials and Methods

2.1 Materials

Three types of plain-woven fabric samples were made using ring spun weft yarns of 30^s Ne with 100% cotton fibres, 50:50% cotton/modal blend and 50:50% cotton/tencel blend yarn in weft. The warp yarn for all fabric samples is cotton/polyester (60:40 with $2/40^s$ Ne). These three fabric samples ware processed for scouring and then dyeing. Thus, a total of nine fabric samples (three samples each of greige, scoured and dyed stage) produced to study the impact of wet processing on properties.

i. **Preparation of yarns**

Weft yarn of 30^s Ne was produced by ring spinning technique using 3.5 twist multiplier. Cotton sample was made by two varieties: J-34 and S-6 in 50:50 ratio. These two varieties are mixed in blow room stage. Modal fibres and tencel fibres were blended each with the cotton fibres in 50:50 ratio at drawing stage. The yarns parameters are given in Table 1.

ii. **Preparation of greige fabric samples**

The plain-woven fabrics were prepared on rapier loom with a warp beam containing cotton/polyester 60:40 (2/40 s count) for all samples. For making fabric samples, three types of yarns: 100% cotton yarn, 50–50% cotton-modal blended yarn and

Table 1 The details of yarn parameters

Description	Cotton	Cotton-modal blend	Cotton-tencel blend
Yarn count	30^s Ne	30^s Ne	30^s Ne
Uster %	9.48	9.56	9.51
Imperfections (−50) (+50) (+200)	0.0 22 47	0.0 37 79	0.0 30 80
Strength RKM CV%	18.83 cN/tex 7.89	20.4 cN/tex 8.7	23.82 cN/tex 9.47
Hairiness	5.39	5.14	5.21
Blend	100% Cotton	50:50	50:50
Blending Technique	–	DF Blending	DF Blending
Yarn TM	3.65	3.65	3.65

50–50% cotton-tencel blended yarns were used in weft only. The ends per/inch and picks/per inch were kept about 61.5 and 43.5 respectively.

iii. **Preparation of scoured fabric samples**

The scouring is done to remove the natural as well as added impurities and make fabric more absorbent. To make scoured samples, all three greige fabric samples were processed in kier for 2 h at boiling temperature using following scouring recipe:

Material to liquor ratio: 1:10.
Caustic soda (alkali): 2% (on the weight of material).
Soda ash: 2%
Detergent: 0.3%
Wetting Agent: 0.1%
Sodium silicate: 0.5%
Sodium sulphite: 0.5%

iv. **Preparation of dyed fabric samples**

Dyed samples were made by dyeing the scoured samples with reactive dye. The dye is able to react chemically with the substrate to form a covalent dye substrate linkage. The reactive dyes have a reactive group and this group makes covalent bond with fibre polymer and act as an integral part of fibre [5].

2.2 *Methods*

The fabric samples were tested for its physical factors (EPI, PPI, Cover Factor, GSM and Thickness), air permeability, moisture management properties, strength and stiffness test.

i. **Physical factors**

Fabric cover factor was calculated using the following formula:

$$\text{Fabric cover factor}(CF_f) = CF_{wrp} + CF_{wft} - \frac{(CF_{wrp} x CF_{wft})}{28}$$

where, CF_{wrp} = Warp cover factor = $n/\sqrt{N}$ (n = ends per inch and N = thread count).

CF_{wft} = Warp cover factor = $n/\sqrt{N}$ (n = picks per inch and N = thread count).

The fabric weight per square meter (GSM) was measured according to ASTM D 3776–02 standards [6].

Fabric thickness was determined at 100 g/cm^2 pressure with the help of a precision thickness tester with accuracy 0.01 mm and pressure foot dia 10 mm.

ii. **Air permeability test**

Air permeability was measured at 185 Pa pressure by using the Digital Air Permeability Tester (YG461E Type) and expressed in mm/s [7] for all samples. The nozzle diameter and test area were 6 mm and 20 cm^2 respectively.

iii. **Moisture management test**

The SDL ATLAS Moisture Management Tester (MMT) was used to measure the liquid moisture management capabilities of the samples [8]. The instrument measures, evaluates and classifies liquid management properties of fabrics. It can measure overall moisture management capacity along with five other properties.

iv. **Fabric breaking strength**

The breaking strength and elongation of textile fabrics may be determined using the strip method test using the Aimil Universal Testing Machine [9].

v. **Fabric stiffness test**

Fabric stiffness is a property of fabric to keep standing without any support. It is an important factor in the study of handle and drape of fabric. The fabric stiffness was measured by Shirley Stiffness Tester [10] work on cantilever principle as per B.S. 3356:1990 test [11]. A 25 mm wide and 200 mm long rectangular strip of fabric was mounted on a horizontal platform in such a way that it hangs like a cantilever and bends downwards. The length of the fabric required to bend to a fixed angle is measured. This is known as bending length.

The stiffness of a fabric in bending is very dependent on its thickness, the thicker the fabric, the stiffer it is if all factors remain same [10]. The bending modulus is independent of the dimension of the strip tested.

$$\text{Bending Modulus} = \frac{12 \times G \times 10^3}{T^3} N/m^2$$

where,

G Flexural rigidity = M X C^3 X 9.807 X10^{-6} μNm,
C Bending length (mm) and dependent on weight of the fabric.
M fabric mass per unit area (GSM).
T Thickness (mm).

3 Result and Discussion

Fabric physical factors (EPI, PPI, Cover factor, GSM and fabric thickness), comfort properties (air permeability and moisture management properties) and mechanical properties (tensile strength, stiffness) are evaluated and results are tabulated in Table 2.

Table 2 Test results

Weft yarn component in fabric	CF	Fabric weight (GSM)	Thickness (mm)	AP (mm/s)	OMMC index	Breaking strength (Newton)	Bending Modulus ($N/m^2 \times 10^{-3}$)
100% C, Gr	17.16	1.29	0.37	117	0.70	185	9.69
50:50 C/M, Gr	17.16	1.28	0.36	145	0.75	166	9.84
50:50, C/T, Gr	17.33	1.28	0.36	129	0.78	196	8.50
100% C, Sc	17.96	1.33	0.39	84	0.81	216	6.62
50:50 C/M, Sc	18.07	1.34	0.38	92	0.78	174	6.31
50:50, C/T, Sc	17.90	1.37	0.38	100	0.79	244	5.86
100% C, Dy	17.80	1.39	0.43	92	0.82	213	6.78
50:50 C/M, Dy	18.22	1.39	0.42	97	0.79	157	6.07
50:50, C/T, Dy	18.19	1.43	0.44	94	0.80	174	4.34

C = Cotton, M = Modal, T = Tencel, Gr. = Greige Stage, Sc. = Scoured, Dy = Dyed, CF = Cover Factor, AP = Air permeability, OMMC = overall moisture management capacity index

3.1 Fabric Physical Properties

i. **Cover factor**

The results of cover factor are tabulated in Table 2 and the impact of wet processing is shown in Fig. 2. Result shows that after scouring, the cover factor increases in all fabrics samples due to fabric shrinkage and further increases slightly in cotton/modal (C/M) and cotton/tencel (C/T) samples after dyeing while in cotton (C) samples, it decreases slightly. There is a very slight difference in cover factor of all the samples.

ii. **Fabric weight (GSM)**

Fabric weight has been analysed and results are tabulated in Table 2 and the impact of wet processing is shown in Fig. 3.

Figure 3 shows that the fabric weight (GSM) of all the three types of fabrics has increased significantly after scouring and further increases in dyeing perhaps due to shrinkage of samples in each step of wet processing.

From the graphical result, it is found that in greige stage, all fabric having nearer same GSM value. After wet processing steps, i.e. in scouring and dyeing, the cotton/tencel blend has highest GSM value followed by cotton/modal blend fabrics and 100% cotton fabric.

iii. **Fabric thickness**

Thickness of a fabric plays important role in its comfort properties as well as handling and drape properties. The data of thickness is tabulated in Table 2 and the impact of wet processing is shown in Fig. 4.

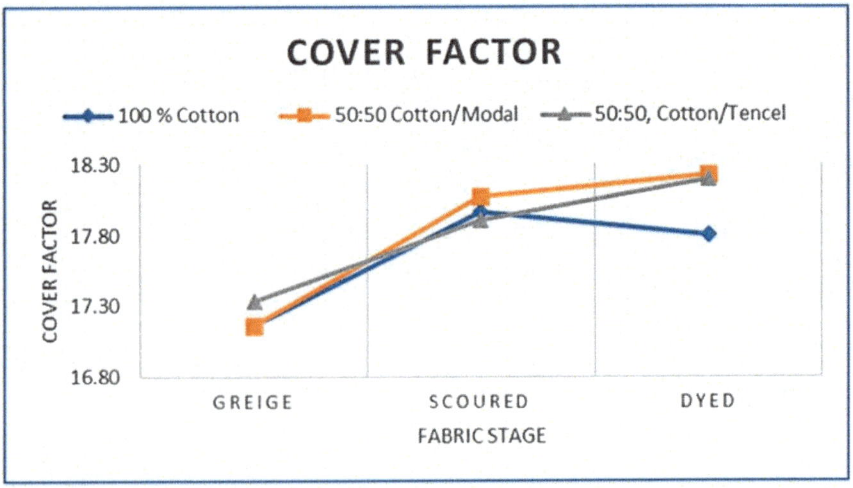

Fig. 2 Fabric cover factor

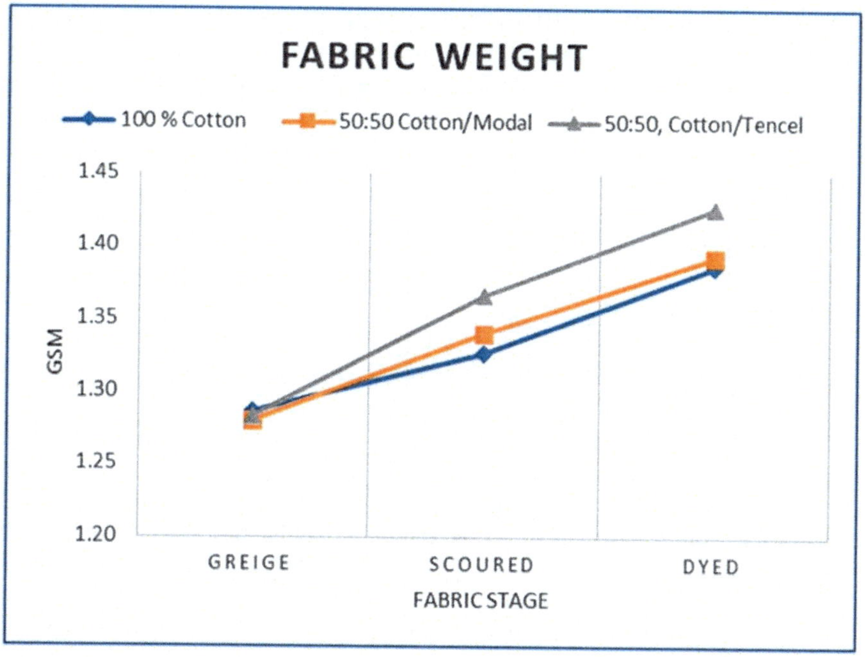

Fig. 3 Fabric weight (GSM)

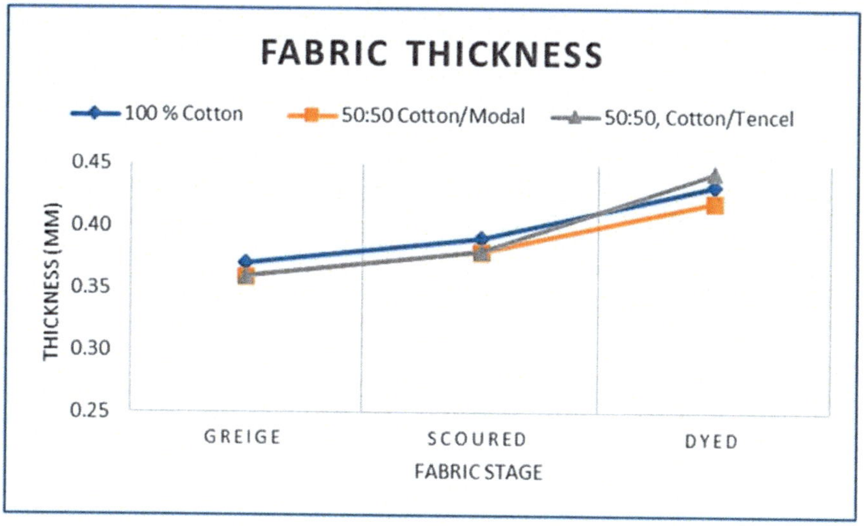

Fig. 4 Fabric thickness

After scoring the fabric thickness increases in all fabric samples significantly and further increases after dyeing. This may be due to shrinkage in process of wet processing. There is a very slight difference in between the different fabric sample in each stage.

3.2 *Comfort Related Properties*

i. **Air permeability**

The results of air permeability of the samples are tabulated in Table 2 and the effects of wet processing on various materials were shown in Fig. 5.

The air permeability decreases after scouring in all samples but there is no any significant changes in after dyeing. This may be perhaps due to mostly all shrinkage takes place at scouring and very little alteration in shrinkage after dyeing. After scouring, the resultant effect of shrinkage and fibre swelling take place and the yarns come closer resulting in decreasing of the gaps to let the air pass and this result in decrease of air permeability after scouring.

In greige stage, the air permeability of cotton-modal blend was the highest followed by cotton-tencel blend and 100% cotton weft yarn fabric samples. After scoring, air permeability of cotton/tencel is highest followed by cotton/modal and 100% cotton fabric. After dyeing, the cotton/modal blended fabric shows highest air permeability followed by tencel/cotton blended fabric and 100% cotton fabric.

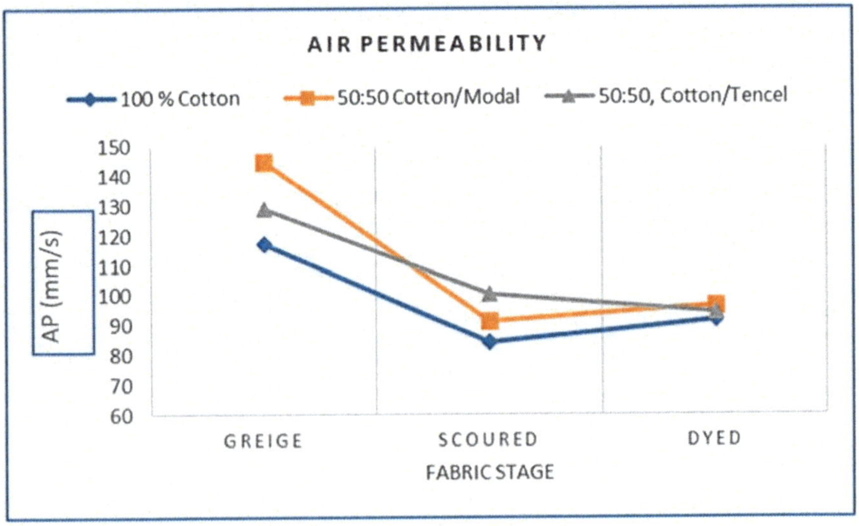

Fig. 5 Air permeability of fabrics

The air permeability of the fabrics depends on structure of the fabric, i.e. size of the pores, fabric thickness and tortuosity of the pores. Most of the air passes through the pores between the yarns but when the fabric is denser, the air also passes through the micro-voids inside the yarns (i.e. the pores between the fibres). There is good correlation (Table 3) between air permeability and fabric cover factor (R = −0.86) and GSM (R = −0.72) and fabric thickness (R = −0.66).

ii. **Moisture management capacity**

The overall moisture management capacity (OMMC) as measured by the Moisture Management Tester are furnished in Table 2 and the effect of wet processing illustrated in Fig. 6.

The overall moisture management capacity (OMMC) is based on three factors, the moisture absorption rate at the bottom of the fabric (ARB), the spreading speed at the bottom (SSB) and accumulative one-way transfer rate of liquid moisture from top to the bottom of the fabric. Higher value of OMMC is associated with higher capillary action.

Table 3 Correlation (R-value)

	Air permeability	OMMC	Breaking strength	Bending modulus
Cover factor	−0.859	0.716	0.016	−0.948
GSM	−0.720	0.651	0.025	−0.906
Thickness	−0.663	0.605	−0.089	−0.753

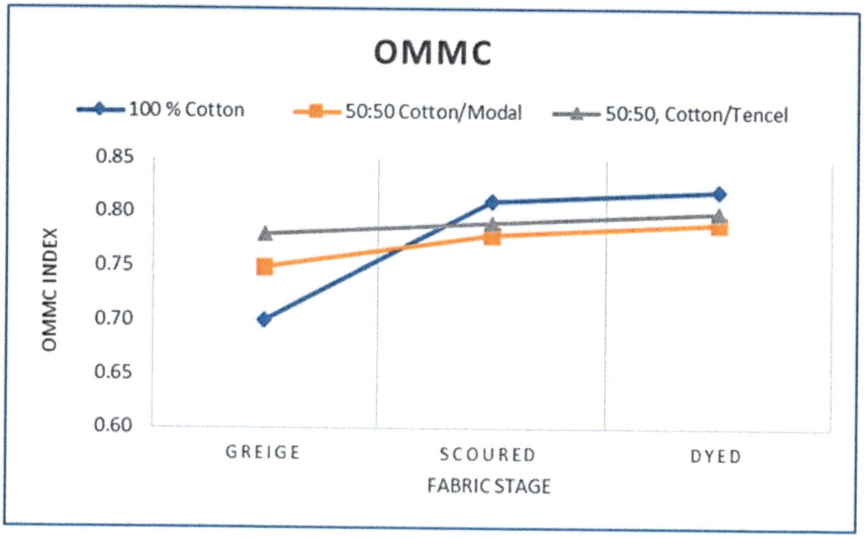

Fig. 6 Overall moisture management capacity

The overall moisture management capacity increases significantly in all samples after scouring and further increases after dyeing. This may be due to removal of cotton wax/spin finish from the cuticle/outer surface of the fibres resulting in improved water transport from the fibrous material. In the greige stage, cotton/tencel blended fabric has the highest value of OMMC followed by cotton/modal blended fabric and 100% cotton fabric. After scouring and dyeing, 100% cotton fabric has highest value of OMMC followed by cotton/tencel and cotton/modal fabric.

There is fair correlation has been found (Table 3) in between OMMC and fabric cover factor (R = -0.72) and GSM (R = 0.65) and fabric thickness (R = 0.61).

3.3 *Fabric's Mechanical Properties*

i. **Breaking strength**

Fabric breaking strength was tested using ravelled strip method test in weft direction and the results are tabulated in Table 2 and Fig. 7.

Figure 7 shows that the fabric strength increases after scouring and decreases after dying in all samples (100% cotton fabric, cotton/modal blended fabric and cotton/tencel blended fabric. In both greige stage and scoured stage, the breaking strength of cotton/tencel blended fabric is highest followed by 100% cotton fabric

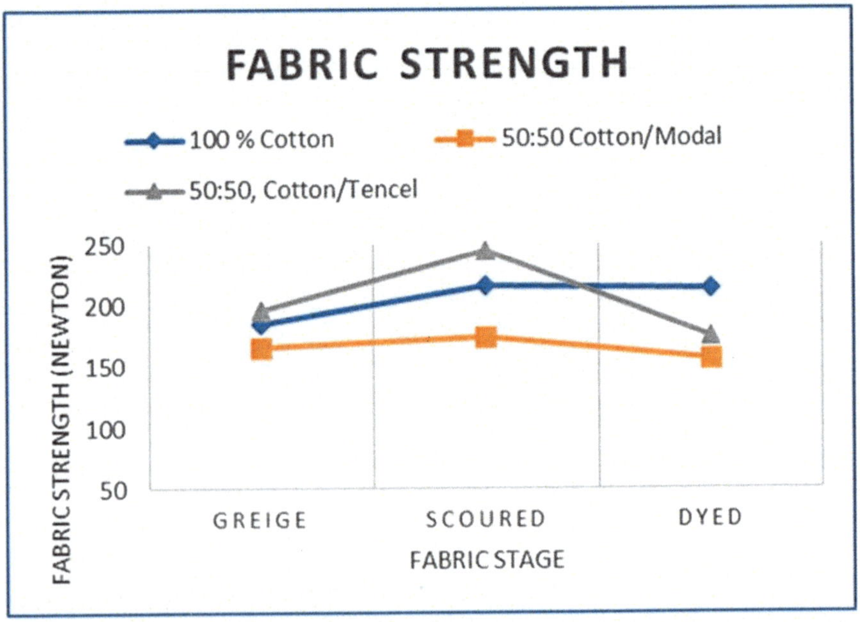

Fig. 7 Fabric breaking strength in weft-wise direction (in Newton)

and cotton/modal blended fabric. More strength of tencel is attributed to crystalline arrangement of its cellulose units which are extremely greatly oriented in the longitudinal axis of the fibre [12]. After dyeing, the breaking strength of 100% cotton fabric is highest followed by cotton/tencel blended fabric and cotton/modal blended fabric. Fabric strength is not good correlated with cover factor, GSM and fabric thickness.

ii. **Stiffness**

Stiffness of fabric was measured in terms of bending modulus. It has impact on fabric handle. As the bending modulus increases, the fabric will be stiffer. More stiff fabric will have less drape thus will not fall freely in a garment. The results are tabulated in Table 2 and the effect of wet processing in weft direction is shown in Fig. 8.

Figure 8 shows the bending modulus of fabrics in various stages, i.e. greige stag, scoured stage and dyed stage predicting the stiffness of samples. The fabric bending modulus decreases after scouring and there is no any significant change has been found after dyeing in mostly all samples.

There is no significant difference that has been found in cotton fabric and cotton/modal blended fabric samples but the cotton/tencel blended fabric has lesser bending modulus in all three stages, i.e. in greige stage, scoured stage and dyed samples.

There is very good correlation (Table 3) of bending modulus with cover factor (R = −0.95), with GSM (R = −0.91) and with thickness (R = −0.75).

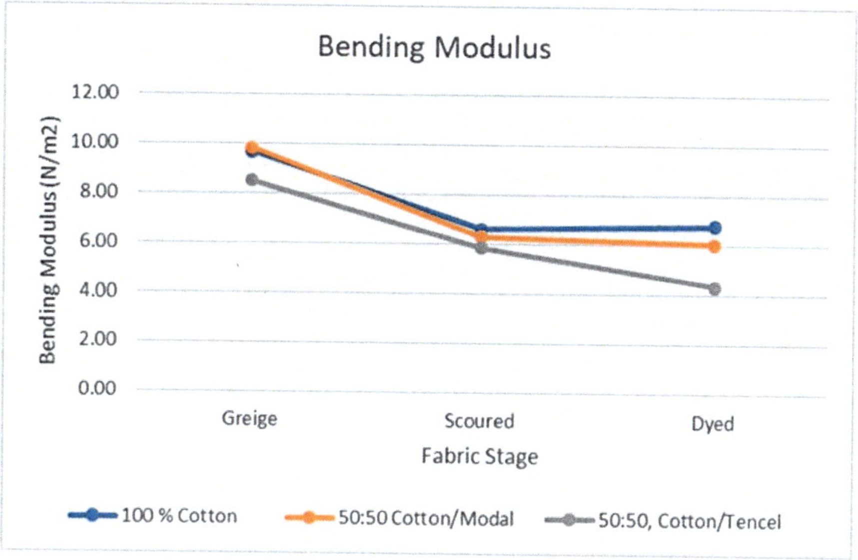

Fig. 8 Bending modulus in weft-wise direction

4 Conclusions

Results are concluded as under:

1. The air permeability decreases significantly after scouring but after dyeing the air permeability alter slightly and this difference is not significant. Air permeability was highest in cotton/modal blend followed by cotton/tencel blend and 100% cotton plain woven fabric in greige stages. After scouring the cotton/tencel blended fabric having the highest value followed by cotton/modal blend and 100% cotton. After dyeing the cotton/modal blend has highest air permeability values followed by cotton/tencel and 100% cotton fabric. In all cases the air permeability of cotton/tencel blend and cotton/modal blend was higher than 100% cotton fabric. There is good correlation between air permeability and fabric cover factor (R = −0.86) and GSM (R = −0.72) and fabric thickness (R = −0.66).
2. After scouring, the OMMC value increases sharply and further increases slightly after dyeing. In greige fabrics, cotton/tencel blend has the highest value of OMMC followed by cotton/modal and 100% cotton fabric. After scouring and dyeing the 100% cotton fabric has highest OMMC followed by cotton/tencel and cotton/modal blend fabric. There is fair correlation has been found in between OMMC and fabric cover factor (R = −0.72) and GSM (R = 0.65) and fabric thickness (R = 0.61).
3. In all fabric samples, the fabric breaking strength increases after scouring and decreases after dyeing. In greige stage and scoured stage, the breaking strength of cotton/tencel blended fabric followed by 100% cotton fabric and cotton/modal blended fabric. After dyeing the 100% cotton fabric shows highest strength followed by cotton/tencel blended fabric and cotton/modal blend fabric. Fabric strength is not good correlated with cover factor, GSM and fabric thickness.
4. In all samples, fabric bending modulus decreases abruptly after scouring and no further change after dyeing significantly. There is no significant difference has been found in cotton fabric and cotton/modal blended fabric samples but the cotton/tencel blended fabric has lesser bending modulus in all three stages, i.e. in greige stage, scoured stage and dyed samples. There is very good correlation of bending modulus with cover factor (R = −0.95), with GSM (R = −0.91) and with thickness (R = −0.75).

References

1. Mukhopadhyay A, Ishtiaque A, Uttam D (2011) Impact of structural variations in hollow yarn on heat and moisture transport properties of fabrics. J. Text. Inst. 102(8):700–712
2. Mather RR, Wardman RH (2011) Regenerated Fibres, The Chemistry of Textile Fibres. RSC Publications, Cambridge, U.K., pp 100–127
3. Udomkichdecha W, Chiarakorn S, Potiaraj P (2002) Relationship between fibrillation behaviour of Lyocell fibres and their physical properties. Text. Res. J. 72(11):939–943

4. Basit A, Latif W, Baig SA, Afzal A (2018) The mechanical and comfort properties of sustainable blended fabrics of Bamboo and with cotton and regenerated fibres. Cloth. And Text. Res. J. 36(4):267–280
5. Dyeing of cotton fabrics with reactive dyes. https://textilelearner.blogspot.com/2012/01/dyeing-of-cotton-fabric-with-reactive.html. Assessed 28 Sep 2019
6. ASTM D3776–96 (Reapproved 2002). Standard Test Method for Mass per unit area (weight) for Fabric, ASTM International, 100 Barr Harbor Drive, PO. Box C700, West Conshohochen, PA19428–2959, United States
7. YG461E/II type, digital air permeability tester. User's Manual, Wuhan Guoliang instrument Co. Ltd., Wuhan, China
8. SDL Atas Moisture Management Tester. https://sdlatalas.com Assessed 25 Aug 2019
9. Aimil Universal Testing Machine (Computerised). https://aimil.com Assessed 25 Aug 2019
10. Saville, B.P.: Objective Evaluation of Fabric Handle. Physical Testing of Textiles. pp. 256–259
11. BS3356:1990. Method for Determining of Bending Length and Flexural Rigidity of Fabrics. British Standard Institute, UK
12. Xu Y, Lu Z, Tang R (2007) Structure and thermal properties of bamboo viscose, tencel and conventional viscose fibre. J. Ther. Ana. Cal. 89(1):197–201

Sustainable Dyeing of Wool by Natural Dyes in Conjunction with Natural Mordants

Neetu Rani and Lalit Jajpura

1 Introduction

Natural colourants have been used in textile, leather as well as food since prehistoric times. These colourants are obtained from natural substances such as animal and vegetable matter with no or very little chemical processing. Synthetic dyes were introduced in 1856 and being cheaper and easily available resulted in a drastic decline in the usage of natural colourants. However, in the present era there has been a revival of interest in natural colourants due to their sustainable behaviour [1, 2]. Environmentalists are always concerned about the use of synthetic colourants in textile industry as they cause waste disposal and water pollution problems [3]. Synthetic dyes and finishes have great environmental concern and thus needs sustainable alternatives [4]. Application of enzymes [5, 6], biopolymers [7–10], herbal finishes [11, 12], natural dyes, and suitable alternatives play a pivotal role in sustainable textile wet processing. Natural dyes do not cause any health hazards being biodegradable hence they can be easily used without much environment concerns. Despite this, use of natural dyes for dyeing textiles has been restricted mainly to cottage industries or at artisan level printers due to associated problem with natural dyes such as lack in reproducibility, poor fastness and cumbersome extraction and application methods [13, 14]. Recently, many commercial printers have started using natural dyes to overcome the environmental damage caused by synthetic dyes. Despite several limitations, there has been a trend to revive the art of natural colouring in recent years due to their distinct

N. Rani (✉)
Department of Fashion Technology, BPSMV, Khanpur Kalan, Sonipat, Haryana, India
e-mail: Neetu.1987@yahoo.co.in

L. Jajpura
Department of Textile Technology, National Institute of Technology, Jalandhar, Jalandhar, Punjab, India
e-mail: jajpural@nitj.ac.in

V. Midha and A. Mukhopadhyay (eds.), *Recent Trends in Traditional and Technical Textiles*, https://doi.org/10.1007/978-981-15-9995-8_12

soothing aesthetic appeal. India being rich in biodiversity has more than 450 plants yielding dyes and pigments for food, textiles and allied industries [15, 16]. However, many of these plant colourants are not yet fully explored for their potential in dyeing textiles. Majority of these plants extracts are being used for medicinal purposes being having good antibacterial properties [16, 17]. Natural dyes along with biopolymers also shown good dyeing and finishing properties [18–20]. The chemical constituents such as quinine, tannin, phenol, etc., present in plant extract provide colour as well as medicinal properties which can be also harvested for producing functional properties to textiles [21]. Thus, dyeing potential of different natural colourants extracted from varied plants, i.e. banyan bark, peepal bark, papaya leaf and Kalanchoe-pinnata leaf were evaluated on wool fabric. Although, only dyeing studies of wool are discussed in this present research paper.

Natural dyes have poor fastness properties hence need mordanting. Pre, meta and post mordanting with heavy metal salts such as aluminium potassium sulphate, ferrous sulphate, copper sulphate, potassium dichromate, etc., is being used traditionally. Although, natural dyes are ecofriendly in nature but owing to use of associated heavy metallic salts in mordanting step makes the dyeing process toxic. Thus, it is required to find out alternatives to heavy metallic salts. Natural mordants can be explored as ecofriendly alternative in dyeing of textiles with natural dyes [22, 23]. Hence, natural biomordants are also used in investigation to compare their behaviour in dyeing with metallic mordants.

2 Materials and Methods

2.1 Materials

2.1.1 Textile Materials

Wool fabric was chosen for this study.

2.1.2 Plant Materials

Four types of plant sources, i.e. banyan bark, peepal bark, papaya leaf and Kalanchoe-pinnata leaf were explored for their dyeing potential.

2.1.3 Mordants

Three chemical mordants such as aluminium potassium sulphate, ferrous sulphate, copper sulphate and four natural mordants such as amla, harda, pomegranate rind and orange peel were taken for this study.

2.2 *Preparation of Dye Solution and Dyeing*

Using optimised conditions of extraction, i.e. MLR 1:30, pH 5 and time 90 min at 100 °C, the dye was extracted from Kalanchoe-pinnata and papaya leaves, peepal and banyan barks [24–26]. It is pertinent to mention that in case of Kalanchoe-pinnata and papaya leaves shade % and mordant % was kept at 20% owing weight of the fabric. Whereas in case of peepal and banyan bark shade % and mordant % was kept at 10% owing to the weight of the fabric based on some preliminary experiments. All fabrics were cut into 20 × 20 cm size samples and dyed for optimising dyeing parameters, i.e. MLR, pH and temperature of the dye bath as well as time of dyeing for obtaining maximum *K*/*S* value. Further, the dyed fabrics were mordanted with four natural mordants such as amla, harda, orange peel and pomegranate to improve fastness properties. Similarly, dyed fabrics with optimised dye recipes were also mordanted with three chemical mordants, i.e. alum, copper sulphate and ferrous sulphate to compare their fastness behaviour with natural mordants.

2.3 *Characterisation of Dye Extract and Dyed Fabrics*

2.3.1 Antioxidant Property

Anti-oxidant property is a measure of the capacity of extracts to scavenge the stable free radicals of DPPH. Samples of 0.20 ml volumes of extracts were added to 3.8 ml of 0.1 mM DPPH solution in ethanol [17]. Samples were put in dark for 30 min to complete the reaction at room temperature for decolourising the solution. Further, decolourisation was assessed on spectrophotometer at 517 nm wavelength and RSA percentage was calculated using Formula (1):

$$\text{Radical Scavenging Activity (\%)} = 1 - \frac{\text{Absorbance (sample)}}{\text{Absorbance (control)}} \times 100 \qquad (1)$$

where $\text{Absorbance}_{\text{sample}}$ refers to the absorbance of the solution having dye extract and $\text{Absorbance}_{\text{control}}$ refers to the absorbance of the solution having the de-ionised water.

2.4 *Anti-microbial Behaviour Evaluation of Dye and Dyed Fabrics*

Extracted dye and dyed fabrics were evaluated for anti-microbial behaviour using AATCC-100 method using gram-positive (*S. aurous*) and gram-negative (*E. coli*) bacteria at IIT, Delhi using quantitative assessment of anti-microbial behaviour.

Equation (2) was used to calculate bacterial reduction percentage.

$$\text{Bacterium Reduction (\%)} = \frac{A - B}{A} \times 100 \quad (2)$$

where, A represents bacterial colonies for the control after 24 h incubation time
B represents bacterial colonies for sample after 24 h incubation time.

2.5 *Analysis of Colour Co-ordinates of Dyed Fabrics*

All the dyed samples were assessed for measuring colour co-ordinates (L, a, b and K/S) using Premier Colour scan computer colour matching system at D65 illuminant/10° observer.

2.6 *Evaluation of Fastness Properties of Dyed Fabrics*

The light fastness, rubbing fastness (wet and dry) and washing fastness of the dyed fabric samples were evaluated as per ISO 105-BO2:2002, ISO-105-X12 and IS: 3361:79, methods respectively.

3 Results and Discussion

Highest K/S value was taken as the optimisation criteria for different dyeing variables and the results are shown in Table 1.

It is clear from Table 1 that all four-extract dyed protein fabric show maximum K/S values in acidic medium.

Table 1 Optimised dyeing conditions for wool fabric four all extracts

Extracts	Variables			
	Time (min)	pH	Temperature (°C)	MLR
Banyan bark	90	3	90	1:30
Peepal bark	90	3	90	1:30
Papaya leaf	90	3	90	1:30
Kalanchoe-pinnata leaf	90	5	90	1:30

3.1 Anti-oxidant Activity of the Extracts

All-natural species of plants have rich phenolics, carotenoids, flavonoids and secondary metabolites in their chemical structure which contribute towards anti-oxidant behaviour. All four extracts were tested for anti-oxidant activity against the free radicals by DPPH because its chemical reaction is very easy to perform [21, 27]. The findings of the study are as follows:

Ascorbic acid calibration curve Eq. (3).

$$y = -0.4892x + 0.7801 \tag{3}$$

where y is absorbance value and x is the amount of Ascorbic Acid.

Resultant value of antioxidant for all extracts is as shown in Table 2.

It can be seen that among all four extracts Kalanchoe-pinnata leaf has maximum Anti-oxidant property so it can be used in cosmetics and finishing of facial wipes etc. for textile application [28].

3.2 Anti-microbial Property

All four extracts and the dyed protein fabric showed very good anti-microbial property against *E. coli* (gram−) and *S. aurous* (gram+) bacteria as mentioned in Table 3. Whereas, papaya leaf extract has maximum bacterial reduction %.

Table 2 Anti-oxidant characteristics of all extracts

Extracts	Antioxidant property	
	Absorbance of extract	Anti-oxidant assay equivalent to ascorbic acid
Banyan bark	0.629	0.308
Peepal bark	0.526	0.519
Papaya leaf	0.523	0.525
Kalanchoe-pinnata leaf	0.445	0.685

Table 3 Bacterial reduction % of all extracts and dyed fabric

Bacteria	Bacterial reduction %							
	Banyan bark		Peepal bark		Papaya leaf		Kalanchoe-pinnata leaf	
	Extract	Fabric	Extract	Fabric	Extract	Fabric	Extract	Fabric
S. aurous	96.06	92.16	96.16	92.79	96.92	94.58	96.82	93.49
E. coli	95.50	90.89	95.63	91.52	96.93	94.19	95.76	92.62

Thus, all these four natural extracts can be used efficiently in medical textiles being excellent antibacterial properties.

3.3 Colour Measurement Using Computer Colour Matching System

The colour co-ordinate values and shades of wool fabric dyed with Banyan and Peepal bark extract in conjunction with various mordants and mordanting techniques are shown in Table 4. Whereas, the colour co-ordinate values and shades of wool fabric dyed with Papaya and Kalanchoe-pinnata leaf extracts in conjunction with various mordants and mordanting techniques are indicated in Table 5.

It can be seen from Tables 4 and 5 that all the dyed samples with different mordants and mordanting techniques exhibit different shades. It can be observed that alum does not have much effect on colour, ferrous sulphate gives tones of grey and copper sulphate changes shades into greenish tone. Pomegranate peel has large amount of tannins hence highlights its own shades in combination with extracts. Harda powder along with extracts modifies the shades of dyed fabrics up to a little extent while Orange peel and Amla powder don't affect the actual shade obtained with true extract. Wool fabric being coarse absorbs large quantity of dye resulting in dark shades. Although, it can be observed that no specific particular trend was observed for any mordant and mordanting technique.

3.4 Colour Fastness Analysis

Tables 6, 7, 8 and 9 show results of colour fastness ratings of the wool dyed fabric with banyan bark, peepal bark, papaya leaf and Kalanchoe-pinnata leaf, respectively. It can be observed from these tables that all the dyed fabrics show satisfactory to good wash fastness. Chemical mordants form H-bond or coordinate bonds with dye and fabric resulting in satisfactory to good wash fastness properties. Although, it can be observed from these tables that rubbing fastness was in generally poor to satisfactory except the Kalanchoe-pinnata dyed wool fabric. Poor rubbing fastness may be occurred due to deposition of natural dyes molecules more on fabric surface instead of penetration inside the interiors of the fabric. Use of appropriate levelling agents may reduce this problem and can improve the rubbing fastness properties.

The results of the study show that all the dyed samples give good light fastness rating or there is increase in darkness of the dyed wool samples instead of fading. The increase in colour depth of some wool dyed sample is due to oxidation of aromatic constituents of natural colourants [29]. It can be also analysed that natural mordants are also comparable to heavy metal-based mordants. Thus, all the four plants extract

Table 4 Colour co-ordinates of wool dyed with banyan and peepal bark extracts

S. No	Name	Wool dyed fabric with banyan bark extract					Wool dyed fabric with peepal bark extract				
		*L**	*a**	*b**	*K/S*	Shade	*L**	*a**	*b**	*K/S*	Shade
Undyed wool		87.8	−1.2	11.7	0.3		87.8	−1.2	11.7	0.3	
W/O mordant		58.1	11.0	19.6	2.5		89.9	21.3	23.19	2.7	
1	Al pre	54.6	13.8	22.6	3.6		55.1	9.7	21.5	3.9	
2	Al meta	62.9	9.4	21.7	2.2		61.8	6.7	21.1	3.6	
3	Al post	59.8	11.4	20.3	2.4		57.8	9.1	20.4	3.1	
4	Cu pre	44.7	10.9	17.3	6.0		48.6	6.0	19.8	5.7	
5	Cu meta	47.2	6.2	18.4	5.7		51.2	3.9	21.8	6.0	
6	Cu post	45.6	8.4	18.2	6.2		47.2	4.6	18.8	6.1	
7	Fe pre	44.3	6.1	10.6	4.6		42.9	3.0	10.0	5.3	

(continued)

Table 4 (continued)

S. No	Name	Wool dyed fabric with banyan bark extract					Wool dyed fabric with peepal bark extract				
		*L**	*a**	*b**	*K*/*S*	Shade	*L**	*a**	*b**	*K*/*S*	Shade
8	Fe meta	42.8	1.3	7.0	4.4		45.8	1.8	10.4	4.9	
9	Fe post	44.0	3.1	9.5	4.5		44.4	2.6	13.1	5.6	
10	H pre	52.3	9.9	25.0	6.2		50.1	8.5	23.2	6.9	
11	H meta	54.6	8.7	26.0	5.8		53.7	7.3	25.7	6.6	
12	H post	57.1	9.0	29.5	5.8		54.9	7.6	26.1	6.0	
13	P pre	50.3	10	23.0	6.0		51.8	8.1	22.6	6.0	
14	P meta	51.3	7.1	23.2	6.4		52.7	7.3	23.8	6.2	
15	P post	54.9	9.8	27.5	6.0		53.5	7.8	24.5	6.0	
16	O pre	51.1	11.1	20.3	4.2		56.1	9.0	19.6	3.5	
17	O meta	53.7	9.2	20.8	4.0		56.4	7.9	20.1	3.4	

(continued)

Table 4 (continued)

S. No	Name	Wool dyed fabric with banyan bark extract					Wool dyed fabric with peepal bark extract				
		*L**	*a**	*b**	*K/S*	Shade	*L**	*a**	*b**	*K/S*	Shade
18	O post	55.7	9.8	22.0	3.7		56.4	8.4	19.9	3.3	
19	A pre	52.8	13.2	19.6	3.5		50.7	8.4	19.4	4.8	
20	A meta	53.9	9.7	18.3	3.3		54.1	7.7	21.0	4.3	
21	A post	57.5	10.6	18.8	2.6		53.6	7.5	20.5	4.3	

Abbreviations Al: alum, Fe: ferrous sulphate, Cu: copper sulphate, H: harda, A: amla, O: orange peel, P: pomegranate peel, Pre: pre-mordanting, Meta: meta-mordanting, Post: post-mordanting

Table 5 Colour co-ordinates of wool dyed with papaya and Kalanchoe-pinnata leaf extracts

S. No	Name	Wool dyed fabric with papaya leaf extract					Wool dyed fabric with Kalanchoe-pinnata leaf extract				
		*L**	*a**	*b**	*K/S*	Shade	*L**	*a**	*b**	*K/S*	Shade
Undyed wool		87.8	−1.2	11.7	0.3		87.8	−1.2	11.7	0.3	
W/O mordant		67.0	2.7	23.9	2.5		66.5	2.8	16.0	1.9	
1	Al pre	62.2	2.7	24.5	3.6		83.8	−1.6	12.2	0.4	
2	Al meta	68.6	1.2	28.2	2.8		72.2	−0.5	19.7	1.6	
3	Al post	68.4	0.3	20.8	1.9		66.7	−1.3	19.2	2.0	
4	Cu pre	49.3	2.5	21.9	6.6		66.4	−5.9	19.4	2.0	
5	Cu meta	50.8	−0.1	24.1	7.0		52.0	−0.4	25.3	6.7	
6	Cu post	52.9	−1.7	22.6	5.7		48.9	1.1	20.1	6.2	
7	Fe pre	52.8	3.6	21.1	5.1		68.2	3.6	16.7	1.4	

(continued)

Table 5 (continued)

S. No	Name	Wool dyed fabric with papaya leaf extract					Wool dyed fabric with Kalanchoe-pinnata leaf extract				
		*L**	*a**	*b**	*K*/*S*	Shade	*L**	*a**	*b**	*K*/*S*	Shade
8	Fe meta	43.2	−0.1	12.9	6.4		44.4	−0.9	8.5	4.9	
9	Fe post	47.8	2.6	18.1	6.0		43.3	0.2	7.3	4.6	
10	H pre	54.1	3.5	26.3	8.0		59.6	4.8	29.2	5.9	
11	H meta	58.7	3.6	27.9	6.5		58.9	4.8	27.6	5.9	
12	H post	57.8	3.5	29.6	7.1		57.9	5.6	30.0	6.9	
13	P pre	51.7	5.0	24.1	7.7		57.1	1.2	25.1	5.9	
14	P meta	56.4	5.7	28.0	7.0		57.7	3.8	28.7	6.4	
15	P post	56.0	6.7	29.3	7.2		59.5	4.1	32.2	7.3	
16	O pre	60.9	1.7	20.8	3.1		79.1	0.5	20.0	0.9	
17	O meta	64.7	2.6	21.5	2.5		67.8	2.6	18.2	1.8	

(continued)

Table 5 (continued)

S. No	Name	Wool dyed fabric with papaya leaf extract					Wool dyed fabric with Kalanchoe-pinnata leaf extract				
		*L**	*a**	*b**	*K/S*	Shade	*L**	*a**	*b**	*K/S*	Shade
18	O post	66.5	2.7	20.0	2.1		68.2	1.8	18.6	1.8	
19	A pre	54.3	3.8	20.0	4.8		60.3	5.2	22.6	3.3	
20	A meta	58.7	4.0	20.6	3.7		62.3	2.9	19.6	3.0	
21	A post	56.4	4.1	18.8	3.7		60.3	3.9	21.9	3.8	

Abbreviations Al: alum, Fe: ferrous sulphate, Cu: copper sulphate, H: harda, A: amla, O: orange peel, P: pomegranate peel, Pre: pre-mordanting, Meta: meta-mordanting, Post: post-mordanting

Table 6 Fastness ratings for dyed wool fabric with banyan bark extract

Applied mordant	Mordanting technique	Wash fastness			Rubbing fastness		Light fastness
		Fading	Staining		Dry	Wet	
			Cotton	Wool			
Without mordant		3/4	4	3/4	3	3	7
Harda	Pre	3/4	4	3/4	1–2	1–2	Darker
	Meta	3/4	3/4	3/4	1–2	1–2	Darker
	Post	3/4	3/4	3/4	1–2	1–2	Darker
Orange peel	Pre	3/4	3/4	3/4	3	2–3	Darker
	Meta	4	4	4	2–3	2	Darker
	Post	4	4	4	3	2–3	Darker
Pomegranate peel	Pre	3/4	3/4	3/4	3	2–3	Darker
	Meta	3/4	3/4	3/4	2–3	2	Darker
	Post	4	4	4	3	2–3	Darker
Amla	Pre	3/4	3/4	3/4	2–3	3	7
	Meta	4	4	4	2–3	2	6
	Post	3/4	3/4	3/4	4	3–4	7
Alum	Pre	4	4	3/4	1–2	1–2	7
	Meta	4/5	4	4	1–2	1–2	7
	Post	4	4	3/4	2	2	7
$CuSO_4$	Pre	3/4	4	3/4	1–2	1–2	7/8
	Meta	4	4	4	1–2	1–2	7/8
	Post	4	4	4	2	2	7/8
$FeSO_4$	Pre	3/4	4	3/4	2	1–2	Darker
	Meta	3/4	3/4	3/4	1–2	1–2	Darker
	Post	3/4	3/4	3/4	1–2	1–2	Darker

Table 7 Fastness ratings for dyed wool fabric with peepal bark extract

Applied mordant	Mordanting technique	Wash fastness			Rubbing fastness		Light fastness
		Fading	Staining		Dry	Wet	
			Cotton	Wool			
Without mordant		4	4	3/4	2–3	2	6
Harda	Pre	4	4	3/4	2–3	2–3	Darker
	Meta	3/4	3/4	3/4	2	2	Darker
	Post	4	4/5	4	2–3	2	Darker
Orange peel	Pre	4	4/5	4	2–3	2–3	7
	Meta	3/4	4	3/4	1–2	1–2	7
	Post	3/4	4	3/4	2	2	7

(continued)

Table 7 (continued)

Applied mordant	Mordanting technique	Wash fastness			Rubbing fastness		Light fastness
		Fading	Staining		Dry	Wet	
			Cotton	Wool			
Pomegranate peel	Pre	3/4	4	3/4	1–2	1–2	Darker
	Meta	4	4/5	4	1–2	1–2	Darker
	Post	4	4/5	4	2	2	Darker
Amla	Pre	4	4/5	4	1–2	1–2	7
	Meta	3/4	4	3/4	1–2	1–2	7
	Post	4	4/5	4	2–3	2–3	7
Alum	Pre	3/4	3/4	3/4	1–2	1–2	6
	Meta	4	4	4	2	2	6
	Post	3/4	3/4	3/4	2–3	2–3	6
$CuSO_4$	Pre	4	4	4	1–2	1–2	6
	Meta	3/4	3/4	3/4	2	2	6
	Post	3/4	3/4	3/4	1–2	1–2	6
$FeSO_4$	Pre	3/4	3/4	3/4	1–2	1–2	6
	Meta	4	4	4	2	2	6
	Post	3/4	3/4	3/4	2–3	2–3	6

Table 8 Fastness ratings for dyed wool fabric with papaya leaf extract

Applied mordant	Mordanting technique	Wash fastness			Rubbing fastness		Light fastness
		Fading	Staining		Dry	Wet	
			Cotton	Wool			
Without mordant		4	4/5	4/5	3	3	7
Harda	Pre	3/4	3/4	3/4	1–2	1–2	Darker
	Meta	3/4	3/4	3/4	1–2	1–2	Darker
	Post	4	4	4	-2	1–2	Darker
Orange peel	Pre	4	4/5	4/5	3	2–3	7/8
	Meta	4/5	4/5	4/5	2–3	2	7/8
	Post	4	4/5	4/5	3	2–3	7/8
Pomegranate peel	Pre	3/4	3/4	3/4	3	2–3	Darker
	Meta	4	4	4	2–3	2	Darker
	Post	3/4	3/4	3/4	3	2–3	Darker
Amla	Pre	3/4	4	3/4	2–3	3	Darker
	Meta	4	4/5	4	2–3	2	Darker
	Post	4	4/5	4	4	3–4	Darker

(continued)

Table 8 (continued)

Applied mordant	Mordanting technique	Wash fastness			Rubbing fastness		Light fastness
		Fading	Staining		Dry	Wet	
			Cotton	Wool			
Alum	Pre	4/5	4/5	4/5	1–2	1–2	7/8
	Meta	4	4	4	1–2	1–2	7/8
	Post	4	4	4	2	2	7/8
$CuSO_4$	Pre	4/5	4/5	4/5	1–2	1–2	7/8
	Meta	4/5	4	4	1–2	1–2	7/8
	Post	4/5	4	4	2	2	7/8
$FeSO_4$	Pre	3/4	3/4	3/4	2	1–2	Darker
	Meta	3	3/4	3/4	1–2	1–2	Darker
	Post	3	3/4	3/4	1–2	1–2	Darker

Table 9 Fastness ratings for dyed wool fabric with Kalanchoe-pinnata leaf extract

Applied mordant	Mordanting technique	Wash fastness			Rubbing fastness		Light fastness
		Fading	Staining		Dry	Wet	
			Cotton	Wool			
Without mordant		4	4/5	4/5	4/5	4	Darker
Harda	Pre	4	4/5	4/5	4/5	3	Darker
	Meta	4/5	4/5	4/5	4/5	3/4	Darker
	Post	4/5	4/5	4/5	4/5	4	Darker
Orange peel	Pre	4	4/5	4/5	4	3/4	No change
	Meta	4	4/5	4/5	4	3/4	Darker
	Post	3/4	4/5	4/5	4/5	4	Darker
Pomegranate peel	Pre	3/4	4/5	4/5	3	2/3	Darker
	Meta	3/4	4/5	3/4	4/5	3/4	Darker
	Post	4	4/5	4	4	3/4	Darker
Amla	Pre	4	4/5	4	4/5	4	No change
	Meta	4	4/5	4/5	4/5	3/4	No change
	Post	3/4	4/5	4	4/5	4	Shade change
Alum	Pre	4	4/5	4/5	4/5	4	No change
	Meta	4	4/5	4/5	4/5	4	Darker
	Post	4	4/5	4/5	4/5	4	Darker

(continued)

Table 9 (continued)

Applied mordant	Mordanting technique	Wash fastness			Rubbing fastness		Light fastness
		Fading	Staining		Dry	Wet	
			Cotton	Wool			
$CuSO_4$	Pre	4	4/5	4/5	4/5	3/4	No change
	Meta	3/4	4/5	4	3/4	3	No change
	Post	4	4/5	4	3	2/3	No change
$FeSO_4$	Pre	3/4	4/5	4/5	4/5	4	Darker
	Meta	3/4	4	4	2/3	1/2	Darker
	Post	3/4	4	3/4	2/3	2	Shade change

as well as natural mordants have very good potential in dyeing and finishing of textiles in an ecofriendly way.

4 Conclusion

All the studied four plant extracts such as banyan bark, peepal bark, papaya leaf and Kalanchoe-pinnata have a very good amount of anti-oxidant contents which make them effective colouring and finishing agents for textiles. Finding of the study shows that all the plant extracts have good affinity towards wool. These plants extracts in conjunction with different natural and chemical mordants give beautiful and wide colour spectrum to wool fabric. It is pertinent to mention that natural mordants showed comparable results of dying and colour fastness to chemical mordants. Hence, natural mordants provide an ecofriendly alternative to toxic heavy metal-based chemical mordants. Besides these, dyed wool fabrics with these four natural extracts possess very high bacterial reduction % leading their application in medical and functional textiles.

In summary, all the four natural extracts in conjunction with natural mordants have good potential in sustainable dyeing of wool fabric with additional antibacterial properties.

References

1. Siva R (2007) Status of natural dyes and dye-yielding plants in India. Curr Sci 92:916–925
2. Ali NF, El-Mohamedy RSR, El-Khatibl EM (2011) Antimicrobial activity of wool fabric dyed with natural dyes. Res J Text Appeal 15:1–10. https://doi.org/10.1108/RJTA-15-03-2011-B001

3. Ali NF, El-Mohamedy RSR, Rajput S (2013) Improvement of antimicrobial activity for onion natural dyed fabrics through chitosan pretreatment. J Appl Sci Res 9:4993–5002
4. Nayak R, Thang Nguyen LV, Panwar T, Jajpura L (2020) Sustainable technologies and processes adapted by fashion brands. In: Nayak R (ed) Sustainable technologies for fashion and textiles. Woodhead Publishing, pp 233–248
5. Jajpura L (2018) Enzyme: a bio catalyst for cleaning up textile and apparel sector. In: Muthu SS (ed) Detox fashion, sustainable chemistry and wet processing. Springer Nature Singapore Pte Ltd, Singapore, pp 95–137
6. Jajpura L (2019) Biotechnology applications in textiles. In: Ul-Islam S, Butola BS (ed) Advanced functional textiles and polymers. Wiley, pp 99–127
7. Jajpura L, Rangi A, Khandual A (2020) Natural finishes, technologies and recent developments. In: Nayak R (ed) Sustainable technologies for fashion and textiles. Woodhead Publishing, pp 209–229
8. Jaipura L, Harad A, Maitra S (2006) Chitin and chitosan in antimicrobial composite fibres. Asian Text J 15:55–58
9. Rangi A, Jajpura L(2017) Guggul gum a biopolymer finishing agent for textiles. In: International conference on textile and clothing-present and future trends (TCPFT-2017), Calcutta, pp 91–94
10. Rangi A, Jajpura L (2015) The biopolymer sericin: extraction and applications. J Text Sci Eng 05:1–5
11. Jajpura L, Saini M, Rangi A (2016) Development of herbal mosquito repellent textiles using essential oils of pine, guggul and rosewood. Colourage 44–48
12. Jajpura L, Saini M, Rangi A, Chhichholia K (2015) A review on mosquito repellent finish for textiles using herbal extracts. Int J Eng Sci Manag 2:16–24
13. Samanta AK, Agarwal P (2009) Application of natural dyes on textiles. Indian J Fibre Text Res 34:384–399
14. Mahesh S, Reddy AHM, Vijaya KG (2011) Studies on antimicrobial textile finish using certain plant natural product. In: International conference on advances in biotechnology and pharmaceutical sciences (ICABPS'2011), Bangkok, 23–24 Dec 2011, pp 253–258
15. Chengaiah B, Rao KM, Kumar KM, Aagusundaram M, Chetty CM (2010) Medicinal importance of natural dyes—a review. Int J PharmTech Res 2:144–154
16. Calis A, Celik GY, Katircioglu H (2009) Antimicrobial effect of natural dyes on some pathogenic bacteria. Afr J Biotechnol 8:291–293
17. Villano D, Fernandez-Pachon MS, Moya ML, Troncoso AM, Parrilla MCG (2007) Radical scavenging ability of polyphenolic compounds towards DPPH free radical. Talanta 230
18. Jajpura L, Paul S, Rangi A (2016a) Sustainable dyeing of cotton with pomegranate rind in conjunction with natural mordant and biopolymer chitosan. Man Made Text India 44:180
19. Jajpura L, Paul S, Rangi A (2016b) Application of chitosan to impart antibacterial property to annatto dyed fabric. Asian Reson 5:99–105
20. Jajpura L, Paul S, Rangi A (2017) Dyeing of cotton with terminalia Chebula as natural dye with chitosan. Asian Dye 14:57–62
21. Pulipati S, Babu PS, Naveena U, Parveen SKR, Nausheen SKS, Sai MTN (2017) Determination of total phenolic, tannin, flavonoid contents and evaluation of antioxidant property of Amaranthus tricolor (L). Int J Pharmacogn Phytochem Res 814–819
22. Rani N, Tanwar P, Jajpura L (2017) Eco fashion by application of tagetes erecta flowers waste on sustainable textile material. Asian Reson 6:31–36
23. Jajpura L, Rimika, Rani N (2018) Sustainable dyeing of cotton fabric with eucalyptus erthrocorys leaves extract and natural mordant. Period Res 6:135–140
24. Rani N, Jajpura L, Butola BS (2017) Extract optimisation for natural herb Kalanchoe-Pinnata. J Basic Appl Eng Res 4:44–48
25. Rangi A, Jajpura L (2017) Effect of MLR, temperature and solvent on the extraction of colouration from arjun (Terminalia Arjuna) bark. J Agroecol Nat Resour Manag 4:87–89
26. Rani N, Jajpura L, Butola BS (2017a) Antimicrobial and antioxidant behavior of natural dye extracted from Kalanchoe pinnata. In: International conference on textile and clothing-present and future trends (TCPFT-2017), Calcutta, pp 94–99

27. Lalas S, Tsaknis J (2002) Extraction and identification of natural antioxidant from the seeds of the *Moringa oleifera* tree variety of Malawi. J Am Oil Chem Soc 677–683
28. Thakare VN, Suralkar AA, Deshpande AD, Naik SR (2010) Stem bark extraction of *Ficus benghalensis* Linn for anti-inflammatory and analgesic activity in animal models. Indian J Exp Biol 39–45
29. Nabawia AA, Moselhey MTH, Guirguis OW (2018) Study the effect of different dyeing conditions of extracted natural dye from leaves of neem on silk fabrics. J Chem Metrol 34–58

Traditional and Technical Assets of Punjabi Culture: Phulkari, Bagh and Chope Embroideries

Maneet Kaur

1 Introduction

Phulkari, Bagh, Chope the legendary Folk embroideries of Punjab have traditionally been an important Article in the bride's trousseau. Phulkari is a vernacular term in Punjabi/Urdu speaking regions of North india (undivided) and it means flower work or crafted floral pattern. Similarly, Bagh is a vernacular term which means garden, used to describe a style of embroidered fabric. Some Baghs were also taken over on the shoulders by brides during lavan (holy sermon to Guru Granth Sahib). Silk floss stitches on hand-spun fabric (khaddar), shimmers on each movement of the bride. The girl used to develop her proficiency in this needlework most certainly after puberty. Her creative endeavours being judged by the number of patterns she could master for each in a systematic and continuous manner. These folk embroideries were usually done for head-covers or wraps on ceremonial occasions. These embroidered wraps were also used as a spread on cot or as back covers of domesticated animals like cows and camels in winters. The records of that time depicts what the common women celebrated, saw, heard or imagined then. Designs and patterns are a translation of what they experienced in day-to-day life, depiction of their possessions, expression of their beliefs and reflection of their imaginative ability.

M. Kaur (✉)
Textile Design, Kala Bhavan, Santiniketan, West Bengal, India
e-mail: 80dr.maneet@gmail.com; mneetdo@gmail.com

Ministry of Culture, New Delhi, India

V. Midha and A. Mukhopadhyay (eds.), *Recent Trends in Traditional and Technical Textiles*, https://doi.org/10.1007/978-981-15-9995-8_13

1.1 Traditional and Technical Identification of Phulkari, Bagh and Chope

Technically, it is a surface embellishment of making design with 'Dabh -taropa' (Darn stitch) in 'Pat' (Silk floss) on 'Khaddar' (hand-spun cotton fabric). Two or three pieces of fabric were sewn together to make a 'Chadder' (wrap). Three of the folk embroideries are visually identified from the appearance of motif, technique and colour combination.

Phulkari: Among three of the traditional folk embroideries one which is much practiced till date is Phulkari. Different motifs are embroidered on the field, border and ends of wrap, making ground fabric visible. Handwoven narrow width, ground fabric was mostly dyed in either red, blue or kept grey (undyed fabric) to embroider in bright, vibrant and contrasting colour with untwisted silk floss (Fig. 1).

Traditionally, Phulkari was made for everyday wear, on which, designs were sparsely embroidered on Khadder. Further Phulkari existed with the above-mentioned basic appearance and has numerous forms, varying from region to region. The significant Phulkari has broadly four types:

1. Sainchi Phulkari,
2. Shishedar Phulkari,
3. Tilpatra Phulkari,

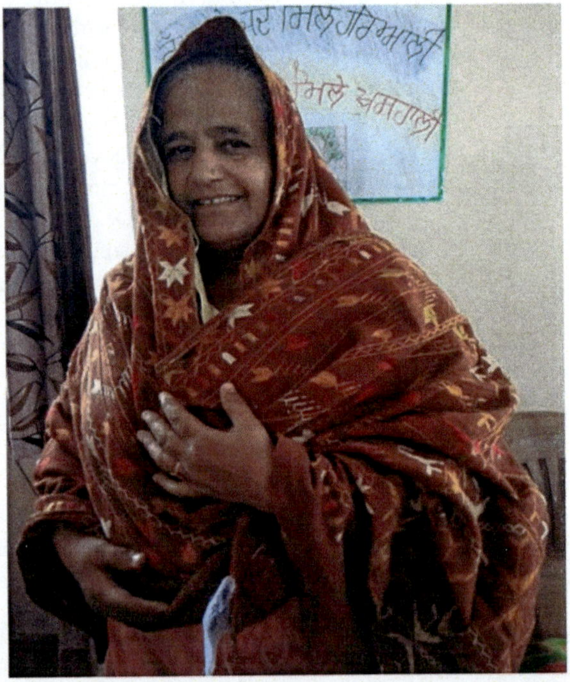

Fig. 1 Ranak Kaur wrapped in Phulkari in Trinjan Mela, organised by Kheti Virasat Mission, Jaito , Punjab

4. Darshandwar Phulkari.

1. *Sainchi Phulkari*: It has figures resembling humans, animals,birds and/or daily use materials were embroidered throughout the field. In this embroidery, outlines of the figures were drawn using black ink (prepared fromcharcoal and kerosene). Some embroiders simply outline-with stem-stich and then fill with long floats of darn stitch. For some auspicious occasionsm motifs of elaborate wedding jewelleries are frequently embroidered on centre of one side of the border , where the shawl falls over the head. Motifs depicted are domestic chores of women such as churning milk, grinding chakki (hand mill), spinning charkha, scene of British officer coming to the village, women carrying umbrellas and walking along with memsahib, scenes such as railways, circus and scenes from popular Punjabi love stories like 'Sohni Mahiwal', 'Sassi-Punnu', were among popular themes. Embroidering household articles and themes like man ploughing, lying on charpoy were also done. Such Chadder wraps are border with blanket stitch and fine phulkari motifs with Chevron,Triangles and geometric pattern.
2. *Sheeshedar*: In this type, fabric has inserts of small square and circular pieces of mirrors which are embroidered with colourful floss of buttonhole stitches holding them in place in order to be part of motif or motif created around them..
3. *Tilpatra*: A tiny circular pieces are primarily embroidered with bands of silk embroidered fabric to add extra embellishment and cover sparsely embroidered motifs.
4. *Darshan Dwar*: An Indigo dyed cotton ground, the name given to this embroidered fabric denotes 'the gate from which one gains a view of the divine, the gates to the Gurdwara or temple are clearly shown on either side'. It was a type of Phulkari which was made as an offering or 'bhent' (presentation) on fulfilment of a wish. It had panelled architectural design. The pillars and top of the gate were filled with latticed geometrical patterns. Sometimes, human beings were also depicted standing at the gate. A counted darning stitch was used to produce a decorative darning. The stitch was worked with the reverse side facing the worker, so that the general design had often been drawn out on the ground fabric with a thread and the worker could follow the design. Usually, each stitch was 1/4th (6 mm) in length and three or four stitches were taken up by the needle at a time (Fig. 2).

Bagh: In Punjabi language, Bagh means garden, wherein the entire surface is embroidered. By working with darning stitch, numerous designs were made by use of horizontal, vertical and diagonal stitches. Such surface embroidered fabric earned name depending upon usage like 'Ghunghat Bagh', 'Vari-da-bagh', 'Bawan Bagh', 'Bhool Bhulaiyaa', 'Parantha Bagh', etc. Culturally, Phulkari carries belief and cultural usage that every Jat Punjabi family should possess an exquisitely embroidered 'Vari-da-Bagh' and as a matter of 'Shagan' (auspicious gesture) should drape the bride on the first time she comes to husband's home (Fig. 3).

Vari-da-Bagh: This shawl envelops the bride when she goes to her new house and her new family welcomes her as part of *vari* or gifts to the bride by her in-laws.

Fig. 2 Phulkari motifs with Nazar Batu motifs in Blue

Fig. 3 Motifs of Bagh, Mrs. Hardeep Kaur, Burj mansa Bathinda

Traditionally, it was started by the grandmother on the birth of a baby boy. Triangle motifs are also found occasionally on the Vari-da-Bagh. This is a special kind of Bagh, which is an example of embroidery with skilful variations of geometrical patterns. Since it is for the bride, it will have some special motifs known as 'Nazar Battu' (a motif or colour to ward away evil eye) inserted in Bagh, the most preferred motif is 'Chirri Chope' Phulkari. It is a bird motif embroidered in the technique of a Chope (Fig. 4).

Bagh as an heirloom is revered and treated as an ancestral blessings by both migrants from Pakistan and residents of Muktsar. Hardeep Kaur, 40 years, Village Mansa Kalan, Bhatinda, was gifted Bagh as 'Shagun' (auspicious gift) at her wedding by her grandmother Gurnam Kaur. This Bagh was embroidered by her grandmother before her marriage at Malookey Fateh village. Later, her Grandmother too taught her Bagh embroidery. At end of this Phulkari piece, the name of her grandfather was written in Punjabi script. According to her lalari (dyer), it was for identification of

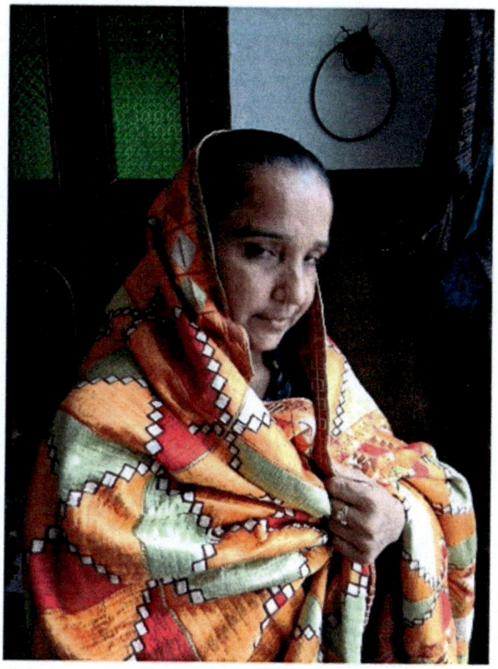

Fig. 4 Hardeep Kaur wrapped in Bagh, embroidered by her grandmother

the owner of the fabric after it had been dyed. She has kept this piece as a token of love by her grandmother and wishes to spread it on the car at her daughter's wedding. It's 3 pieces sewn fabric in which two pieces are 21 inches and the central one is 101/2 inches.

Chope: A folk embroidered fabric with ground fabric in red and embroidered with silk floss in yellow. Traditionally, two narrow, handwoven fabric panels were sewn that were embroidered in similar patterns on both sides. The only motif embroidered on both selvedge was a series of triangles with the base towards the selvedge and pointing inwards. The design was worked with small squares in a step-ladder fashion. Chope Phulkari is made with a double running stitch so that it is identical on both sides of the fabric. The finest use of silk floss in embroidery is reflected in Chope. Rarely this style of folk embroidery is found to be in practice or in use (Fig. 5). Technically, this type of embroidery shares common stitches of Karnataka embroidery called Kasuti.

Fig. 5 Chope motifs embroidered with yellow Silk floss

2 Common Technical Aspect

2.1 Materials-Ground Fabric, Threads, Dyes, Motifs, Auspicious Motif in Phulkari, Bagh, and Chope

1. *Ground/Base fabric*: The plains of Punjab are rich cotton-producing areas and every nearby village has a settlement of weavers. Handspun Khaddar was made of yarn spun from cotton fibers by the women either by themselves or by village weaver on the charkha (spinning wheel). After making, the yarn was woven in plain weave and was dyed by the 'Lalaris' (8). Spinning Cotton fibers to yarns was household occupation (5), for some females weaving was also homecraft.So very few used to depend on lalaris for plain woven fabric and on Rang-rez (dyer) for desired colour on fabric. Ground fabric colour was always 'Nabhi' (tones of reds) and every colour looked good on this ground colour. Other ground colours such as black, brown, blue and even green were also used for making Phulkari but to a lesser extent. Name of one of the family members was used to be written on end of fabric before given that to 'Rang-rez' (dyer) (8). Locally available, cheap, hard-wearing and most importantly, it was preferred to mill-made cloth, as it is coarse weave facilitated the counting of threads necessary for Phulkari work. Three to four narrow loom width of cloth were joined to form complete shawl, 17–23 inches width (4).

2. *Threads*: In the context of traditional threads, two types of threads were used, pat (untwisted silken floss) and white cotton yarn. Pat was considered a 'Suchha Dhaga' (pure thread). It was available in the form of 'lachhiyan/guttiyan' (Skeins) and being an expensive yarn, was sold per tola (traditional South Asian unit of mass). The yarn was usually bought in bulk to embroider many Phulkaris or the silk floss (pat) had been in use in various embroideries like suf of Gujarat, Turkish folk art of Swat valley in Pakistan, Chamba Rumal in Himachal Pradesh. The untwisted silk skeins, used as the embroidery thread, were imported from Afghanistan, Kashmir, and Bengal and sold by itinerant merchants by weight to the hawkers or peddlers, who sold things of daily needs from village to village (8).These untwisted silk floss give smooth flow on Khadder, very fragile and entwines in characters and mostly long stands are inuse and held pulled.
3. *Dyes*: Ground fabric used to be naturally dyed with Kikkar (Babul bark) and Manjistha (indian Madder),so shade used to be deep brown or red. Silk floss of Phulkari/Bagh/Chope are folded reverse side out and carefully washed with salt or vinegar in water knowing that the colours will bleed in water if washed in hard detergent. Madder red dye or indigo for ground fabric, silk floss, once dyed at Dera Ghazi Khan (now in Pakistan), Amritsar or Jammu. Vernacular names for colours are: 'Khatta' (Yellow/Orange), 'Gulabi' (Pink), 'Hara/Angoori' (Green) in 'Pat' (Flat untwisted Silk floss) and 'Sucha' (White cotton hand-spun yarn).
4. *Motifs*: They are composed in grid or linear arrangement as technique follows darn stitches over counting from reverse side. Motifs are without references from books or tracing of Phulkari motifs, thus developed from mere imagination and calculation. Traditionally, the use of coarse Khaddar fabric made it easy to count the yarn. The hallmark of Phulkari is making innumerable patterns by using long and short darn stitches. The designs, geometrical are usually on a Khadi, Khaddar (Coarse cotton cloth) stitched with 'Heer' (Silk floss) thread. The intricate counted double stitch is seen in Chope Phulkari. The journey with the stitch has to be known before stitching starts. . The designs , techniques and patterns were not documented but transmitted from word of mouth and each regional group was identified with the style of embroidery or design. Embroidered in soft untwisted floss silk, colours in combinations of gold, yellow, white, orange or red, on the ground that was usually a brick red colour but could sometimes be blue, white. The first colour to be filled by embroidery on the ground fabric was especially yellow and other colours then followed. The amount of white yarn was less in comparison with the coloured pat (silk floss) and filled in the end as it used to get dirty easily. There were Phulkaris/Baghs with one single colour, preferably red pat on white ground fabric called Thirma.The technique of maintaining relation of thread and fabric gives immense freedom of expression to develop motifs withharmony in blending of colour and vibrancy of patterns in making exclusive and unique.
5. *Method to stitch*: Specially for Bagh, embroider sits on the ground and the working fabric is kept between knees. From right hand, embroidery is carried on top of the knee and the left hand maintains stretch of fabric, in a way motif with silk floss gets aligned and regular pressure is given. Only one thread was

Fig. 6 Handwoven narrow width fabric sewn in runing stitch to make broad wrap

taken up with each pick of the needle, leaving a long stitch below to form the pattern. Stitching ran in both horizontal and vertical directions in order to give a variation in texture. Phulkari was once embroidered without a frame as the ground fabric was coarse. It helps to retain the reflection of silk back to eyes while embroidering. The stillness and concentration involved in embroidering a needle, matching colours and simply watching a pattern grow helps beat stress. Reverse side embroidery was preferred for the technical reason to avoid the striking reflective glare of silk glossy thread.

6. *Nazar Battu* (auspicious motif, to ward away evil eye): These are created in the dark shade of silk floss either on an ongoing pattern or on a new pattern with similar outlook, e.g., blue triangles, blue lines, chiri chop motif at the end of the fabric. Only one flat strand of silk floss was taken up with each pick of the needle, leaving a long stitch below to form the pattern. Stitching ran in both horizontal and vertical directions in order to give variation in texture. It could be embroidered without a frame as it did not pucker or pull (Fig. 6).

3 Emerging Trends in Terms of Material, Method, Patterns and Usage of Phulkari, Bagh and Chope

Due to migration after partition and growing of the nuclear family system, the culture of making homemade Phulkari has turned to commercial made Phulkari. The demand for Phulkari still continues, but to meet that, some artisans are skilled to follow a pattern without knowing the cultural significance and technical method of making motif. Contemporary embroidery follows accuracy on making meticulous Phulkari/Bagh motifs and put more labour on achieving nearest visual similarity. Skies of silk floss were dyed at home with acid dyes and now are replaced by rayon

Fig. 7 Bagh embroidered by my Great Grandmother presented for talk at NIT, Conference

viscose flat untwisted mill dyed yarns that come from Delhi, Ludhiana and Mumbai. Similarly, new ground fabric is introduced that is available in 45 inches width in fine Cotton, Chiffon, Crepe. The symmetrical linear quality of Darn stitch has either combined with other types of embroideries or emerged as a single motif repeated as all-over patterns from machine embroidery for the wraps. With the growing tendency of Photogenic Cultural presentations, Phulkari/Bagh from wraps has entered into accessories, upholstery and dress materials (Fig. 7).

Acknowledgements Being from a Punjabi family and fourth in the generation to practice the craft of Phulkari/Bagh/Chope was a challenge in terms of identification of authentic Phulkari. Except one Bagh Chhader(wrap), which is my paternal family heirloom, no other sources of living with traditional materials were there. Gradually, as a faculty in the Design Department of Apeejay College of Fine Arts, I got an opportunity to develop skills and learn from local sources, elderly peple like Kawaljeet kaur, 70+ old, and later joined team of collegues and Principal Dr.Sucharita Sharma, to understand, learn and further prepared students for Phulkari making competition in zonal youth festival. Further, with the support and guidance of my big Family especially parents Amarjeet Singh Chhatwal and Pritpal Kaur and their friends Husband Ameet Singh , my Friends especially Neha Bansal and Amar Bansal, Charanjeet Narula , Teachers especially Baudev Biswas, Mentor Late Y.K.Sinha and my inquisitive students developed my own style of creating fine art pieces in Phulkari that led to being a fellow of CCRT fellowship by the Ministry of Culture India. I am grateful to Chairman Professor Vinay Midha, Convenor Dr. Monika Sikka , Jury member Dr.Mojgan Jahanara and their team for accepting my paper for 2nd International Conference on Emerging Trends in Traditional and Technical Textiles from November 1–3, 2019, NIT Jalandhar, Punjab (INDIA).

References

1. Chattopadhyay K (1977) Indian embroidery. Wiley Eastern Limited, Safdar-jang, Enclave, New Delhi
2. Dhamija J (2004) Asian embroidery. Crafts Council of India, Shakti Malik, Ab-hinav Publications, Hauz Khas, New Delhi
3. Fisher, N (1995) Mud mirror and thread: folk traditions of rural India. Maplin Publishing Pvt. Ltd, Ahmedabad in association with the Museum of New Mexico, Santa fe, Mexico

4. Gillow J, Barnard N (2008) Indian Textiles,OM Books International, Ansari Road, Darya Ganj, New Delhi
5. Graham J, Dhamija J (2004) Asian embroidery, Indian sub continent: Phulkari and Bagh: The embroidery shawls of the Punjab, Abhinav Publication. ISBN: 8170174503
6. Hitkari SS (2003) Designs and patterns in Phulkari. Phulkari Publications, 8, Birbal road, Jangpura, Extensions, New Delhi
7. Irwin J (1973) Margaret Hal, Indian embroideries, vol 2. Historic Textiles of India at the Calico Museum, Published by S. R. Bastikar, Ahmedabad
8. Kaur J, The past and present of Phulkari, The Tribune. Updated 6 Mar 2015
9. Malik SB (2011) From silk to synthetic Phulkari: the long journey of period textiles. Int J Human Soc Sci 1(16)
10. Lehri RM, Khan JM (2006) Indian embroidery, ethnic and beyond. Super book house Sind Chambers, Colaba, Mumbai, 400005, India
11. Morrell A (2008) Indian embroidery techniques at the Calico Museum of Textiles; a working guide. Sarabhai Foundation, Opp. Underbridge, Shahibaug, Ahmedabad
12. Paine S (2001) Embroidery from India and Pakistan. The British Museum Press, Bloomsbury Street, London
13. Singh KK (1998) Punjab, art and culture. Atma Ram and Sons, Kashmere Gate, Delhi, Lucknow
14. Shrikant U (2008) Ethnic embroidery of India. Samata Enterprises, King's Apartment, Juhu Scheme, Mumbai
15. Kaur A, Kallan A (RIMT University, Mandi Gobindgarh, Punjab, India) The threads of emotions-Phulkari. Int J Trend Sci Res Dev (IJTSRD), Int Open Access J 2(4). https://www.ijtsrd.com/. Last accessed 12 May 2019. IISNNo: 2456-6470
16. Kaur P, Revival of Punjab's traditional handicraft: Phulkari; Asian J Manag. ISSN-0976-495X www.anvpublication.org; https://www.researchgate.net/publication/303844447_Revival_of_Punjab's_Traditional_Handicraft_Phulkari. Last Accessed 13 July 2019
17. Aulakh TA Culture history : Phulkari and Bagh, Academy of the Punjab in North America, A non-religious and non-political organization of all Punjabies for Pro-motion of Punjabi language, Literature and Culture. http://apnaorg.com/prose-content/english-articles/page-62/article-3/index.html. Last accessed 18 July 2019
18. Aziz A, Phulkari Baghs of the Punjab: another perspective. TAASA review, vol 26(3). https://eprints.soas.ac.uk/25078/1/aziz-phulkari-baghs-article-TAASA-2017.pdf. Last accessed 20 July 2019
19. Walia N, The Times of India, 'Phulkari: A Stitch in Time', 1 April 2018. https://Timesofindia.indiatimes.com/lifestyle/fashion/buzz/a-stitch-in-time/articleshowprint/63531004.cms. Last accessed 19 July 2019

Optimization of Fusing Process Conditions Using the Response Surface Design

Ashish Hulle and Ravikumar Purohit

1 Introduction

Interlining is a layer of fabric placed between the shell fabric and facing. One side of the interlining fabric has thermoplastic resin, which on heating is responsible for joining shell fabric and facing. Interlinings assume a significant job fit as a fiddle into the detail zones of garments, for example, the fronts of coats, collars, lapels, sleeves, and pocket folds [1]. Additionally, they settle and fortify zones subject to additional wearing pressure, for example, neck areas, facings, fix pockets, belts, plackets, and catch gaps [2, 3]. Elements of fusible interlining in articles of clothing can be abridged as the simplicity of piece of clothing fabricating because of dependability of shell texture, blessing of volume because of good formability, outline and shape maintenance of piece of clothing because of reiteration of cleaning, accomplishment of fitting adaptability, improvement of the look, fall and material properties of a delivered piece of clothing [2, 3]. Although interlining is an undetectable inside piece of an article of clothing, the interlining development and the combination procedure of interlining and shell texture influence sewability, appearance, toughness, handle, and mechanical properties of the article of clothing [4, 5]. A decent fused textile material can be obtained when a privilege fusible interlining is picked for a given fabric and when ideal fusing conditions (fusing time, temperature, and pressure) are resolved. Choosing a correct sort of fusible interlining is still to a great extent dependent on experimentation strategy just as experience. Henceforth, it is worth to study the impact of fusing time and temperature on qualities of fused fabric.

A. Hulle (✉) · R. Purohit
Department of Textiles, D.K.T.E. Society's Textile and Engineering Institute, Ichalkaranji 416115, India
e-mail: ashishhulle@hotmail.com

V. Midha and A. Mukhopadhyay (eds.), *Recent Trends in Traditional and Technical Textiles*, https://doi.org/10.1007/978-981-15-9995-8_14

Table 1 Properties of fusible interlining

S. no	Specification	Value
01	Composition	100% terylene
02	Coating	PA
03	Base fabric	Plain
04	Glue type	Double dot, small
05	Areal density	200 gm/m^2
06	Thickness	0.35 mm
07	Tensile strength	53 gmf
08	Elongation	26.8%
09	Bending modulus	341.06 kg/cm^2
10	Color	White

Table 2 Properties of men's shirting fabric

S. no	Specification	Value
01	EPI	60
02	PPI	46
03	Areal density (g/m^2)	160
04	Thickness (mm)	0.35
05	Warp way tensile strength (gmf)	64
06	Warp way elongation (%)	20
07	Weft way tensile strength (gmf)	35
08	Weft way elongation (%)	27.5
09	Bending modulus (kg/cm^2)	76.55
10	Hot air shrinkage (%)	1.2
11	Hot water shrinkage (%)	2.01

2 Experimental

2.1 Material

Woven fusible interlining with polyamide coating was fused on commercially available men's cotton shirting fabric. Properties of fusible interlining and men's shirting fabric are represented in Tables 1 and 2, respectively.

2.2 Methodology

Woven fusible interlining was fused on commercially available men's cotton shirting fabric by varying fusing conditions like fusing time and temperature at constant

Table 3 Sample runs

Run/sample no	Time (s)	Temperature (°C)
S1	10	120
S2	10	130
S3	12	120
S4	12	130
S5	14	120
S6	14	130
S7	16	120
S8	16	130

fusing pressure of (1.5 bar). Fusing was carried out at two different temperatures, viz., (120°C, 130°C) with four different fusing time (10 s, 12 s, 14 s, 16 s). The experiment was carried out with two factors, viz., fusing time having 4 levels and fusing temperature having 2 levels, as shown in Table 3.

This orthogonal array is chosen due to its capability to analyze the interactions among factors. Fusing parameters were standardized after preliminary experiments. Selection of the levels is based on the interlinings fusing parameters specifications. The fusing process was carried out in factory conditions, on continuous fusing press.

Sixteen fused samples were produced at each fusing condition. Out of 16 samples, 8 samples were subjected to the dry-cleaning process. Dry cleaning was done using tetrachloroethylene, which is the most widely used solvent for commercial dry cleaning. The solvent temperature was maintained at 30 °C. After cleaning, samples were tumbled in a stream of warm air.

2.3 *Testing*

All specimens were subjected to bending rigidity for fusible interlining, bending rigidity for fused fabric, bond strength, tensile strength, elongation, and shrinkage tests. Details of all these tests are shown in Table 4.

3 Results and Discussion

This investigation was divided into two parts. In the first part effects of fusing conditions on properties of men's shirting fabric has been identified and in second part optimization of fusing process conditions using the response surface design has been done.

Table 4 Details of tests

Run/sample no	Time (s)	Temperature (°C)
01	Areal density	GSM cutter
02	Thickness	Thickness tester
03	Bending length	Stiffness tester
04	Bond strength	Instron-5565
05	Tensile strength and elongation	Instron-5565
06	Hot air shrinkage	Oven
07	Hot water shrinkage	Boiling water pan

3.1 *Effect of Fusing Conditions*

Bending Modulus

Bending modulus is the key indicator stiffness of material irrespective of the type of material and its characteristics. After determining the bending length bending modulus of all the specimens had been determined and the results are tabulated in Table 5. Also, the effect of fusing condition on bending characteristics of the normal and dry-cleaned specimen is represented in Fig. 1.

From Fig. 1, one can depict that fusing time and fusing temperature are directly proportional to bending modulus. Increase in fusing time and fusing temperature significantly increases the fabric stiffness increasing the fabric stability.

This trend is because of increased bending length due to increase in fusing time and temperature. During the process of fusing, because of longer fusing time and higher fusing temperature, the resin melts properly and spreads over the fabric causing the adhesion of the interlining with fabric. As compared to the unfused fabric on fusing bending modulus increased by 568%. Almost 5–6 times rise in bending modulus represents the higher shape retention characteristics of fused fabric. The dry-cleaning

Table 5 Bending modulus (kg/cm^2)

S. no	Time (s)	Temperature (°C)	Bending modulus (kg/cm^2)	
			Normal	Dry cleaned
S1	10	120	6.8	6.9
S2	10	130	7.3	7.0
S3	12	120	7.4	7.3
S4	12	130	7.6	7.5
S5	14	120	7.7	7.7
S6	14	130	7.7	7.7
S7	16	120	7.9	7.8
S8	16	130	8.0	7.8

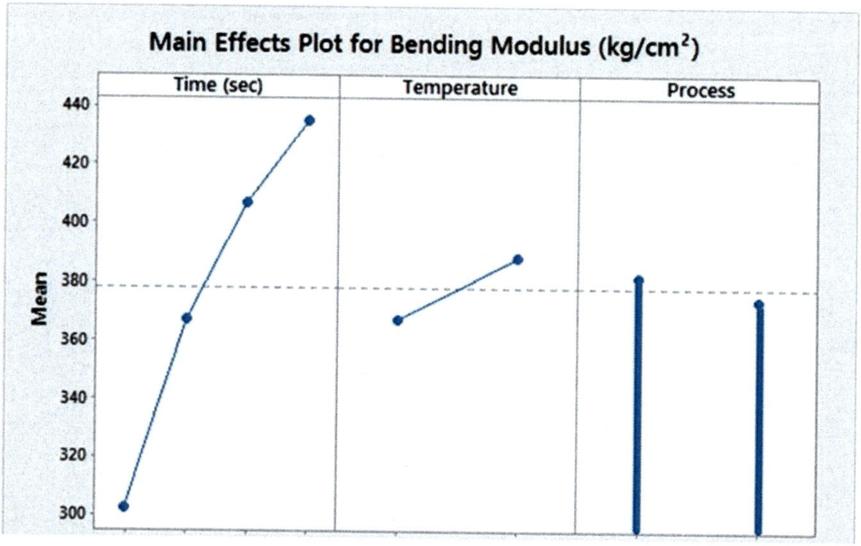

Fig. 1 Bending modulus (kg/cm^2)

process does not affect the fabric stiffness. Even though there is a decrease in bending modulus after dry cleaning, this change is insignificant. S8 represents higher bending modulus in all cases.

Bond Strength

Bond strength of fused systems is one of the most important parameters, which greatly influence the quality of a garment. The strength of the connection depends on the adhesive forces between the fabric and the adhesive. Table 6 represents the mean values of bond strength of fused fabric in both directions (warp and weft) before and after dry cleaning.

Table 6 Bond strength (kgf)

S. no	Time (s)	Temperature (°C)	Normal		Dry cleaned	
			Warp way	Weft way	Warp way	Weft way
S1	10	120	13.76	13.93	13.47	13.64
S2	10	130	14.74	14.15	14.43	13.85
S3	12	120	16.48	16.32	16.13	15.98
S4	12	130	16.87	16.82	16.52	16.47
S5	14	120	18.37	18.65	17.98	18.29
S6	14	130	18.34	18.76	17.95	18.37
S7	16	120	21.48	21.48	21.03	21.03
S8	16	130	21.76	21.53	21.30	21.08

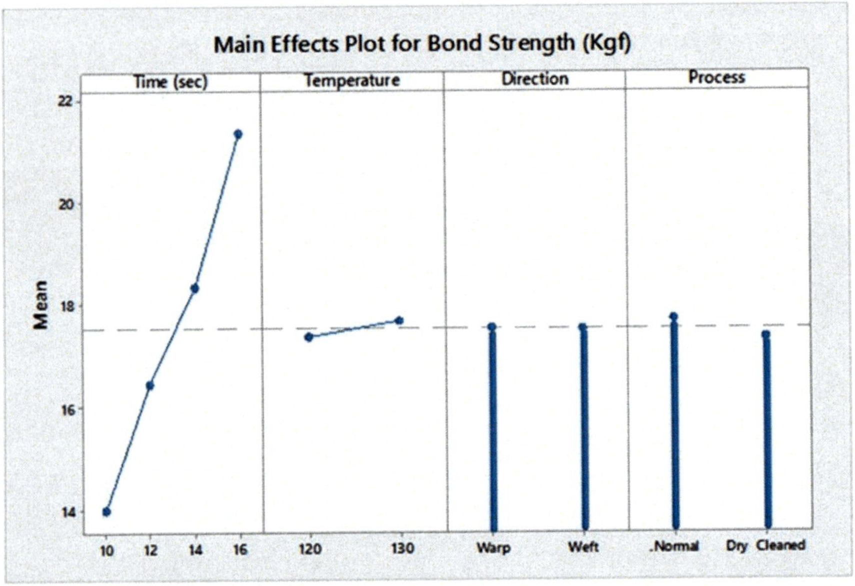

Fig. 2 Bond strength (kgf)

From Fig. 2, it can be clearly observed that there is a significant increase in bond strength as we increase the fusing time and temperature. During the process of fusing, it has been observed that the adhesive or the resin that is present on the surface of interlining will melt and flow into the interstices of the fabric thus enabling a deeper and a bigger contact area. Therefore, more energy is required in the breaking of bonds and hence the bond strength is high. This rate of melting and flow increases with increase in fusing time and temperature.

In the case of bond strength, directional effects are negligible. Irrespective of other factors all fabric samples show same band strength as far as warp and weft direction is concerned. The fabric onto which the interlining is fused is made up of the same yarns (fibers) in both warp and weft direction. Similar kind of material in both warp and weft direction have the same bonding efficiency in either direction. Hence bond strength is the same in the warp as well as in weft direction. Though there is a decrease in bond strength after dry cleaning, the effect is not significant. On multiple washing cycles, there may be a significant reduction in bond strength. This can be attributed to the poor stability of adhesive in dry cleaning solvents. But this needs further investigation, which is out of the scope of the current topic.

Tensile Strength and Elongation

Tensile strength is an important property as it relates to the strength and performance of the material. The results of the measurements of tensile strength and elongation values of fused specimens are given in Tables 7 and 8, respectively. Effect of selected

Table 7 Tensile strength (kgf)

S. no	Time (s)	Temperature (°C)	Normal		Dry cleaned	
			Warp way	Weft way	Warp way	Weft way
S1	10	120	70.20	69.35	68.80	67.96
S2	10	130	89.70	78.26	87.91	76.69
S3	12	120	87.75	74.30	86.00	72.81
S4	12	130	75.40	92.12	73.89	90.28
S5	14	120	101.40	94.10	99.37	92.22
S6	14	130	116.35	99.05	114.02	97.07
S7	16	120	111.80	104.00	109.56	101.92
S8	16	130	124.80	109.94	122.30	107.74

Table 8 Elongation (%)

S. no	Time (s)	Temperature (°C)	Normal		Dry cleaned	
			Warp way	Weft way	Warp way	Weft way
S1	10	120	32.45	33.21	33.10	33.87
S2	10	130	30.40	31.11	31.01	31.73
S3	12	120	31.52	32.26	32.15	32.90
S4	12	130	29.40	30.09	29.99	30.69
S5	14	120	28.25	28.91	28.82	29.49
S6	14	130	27.60	28.24	28.15	28.81
S7	16	120	26.34	26.95	26.87	27.49
S8	16	130	26.04	26.65	26.56	27.18

factors on tensile strength and elongation of fused fabric has been represented in Figs. 3 and 4, respectively.

Among the fused samples, S8 has the highest tensile strength compared to other samples. As compared to unfused sample, the tensile strength of the fused sample in warp direction increases by 194% and by 311% in the weft direction. These results are enough to explain the reinforcing effect of interlining on fabric.

When the load is applied on the fabric, initially the load is carried by both interlining and fabric. When applied load reaches maximum bond strength or holding, the power of interlining polymer, delamination starts. Because of delamination, there is a nonuniform distribution of load on both on interlining and fabric. Further loading results in rupture of either interlining or fabric. Then there is a fracture in fused fabric. Therefore, the load required to cause breakage of fused fabric is high compared to that of unfused fabric.

As fusing time and temperature increases, the tensile strength of fused specimen increases. The same reason can be given for these trends as given for bond strength. Warp way tensile strength is significantly higher than that of weft way strength.

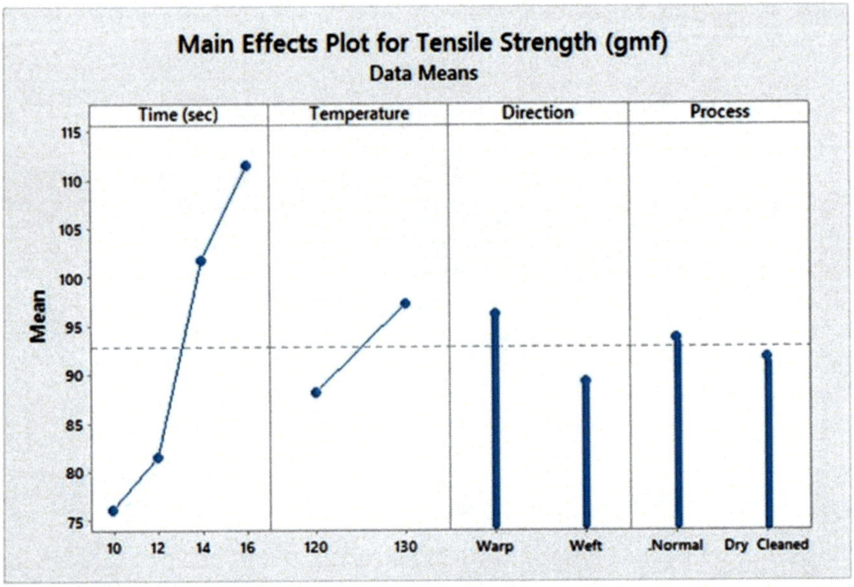

Fig. 3 Tensile strength (kgf)

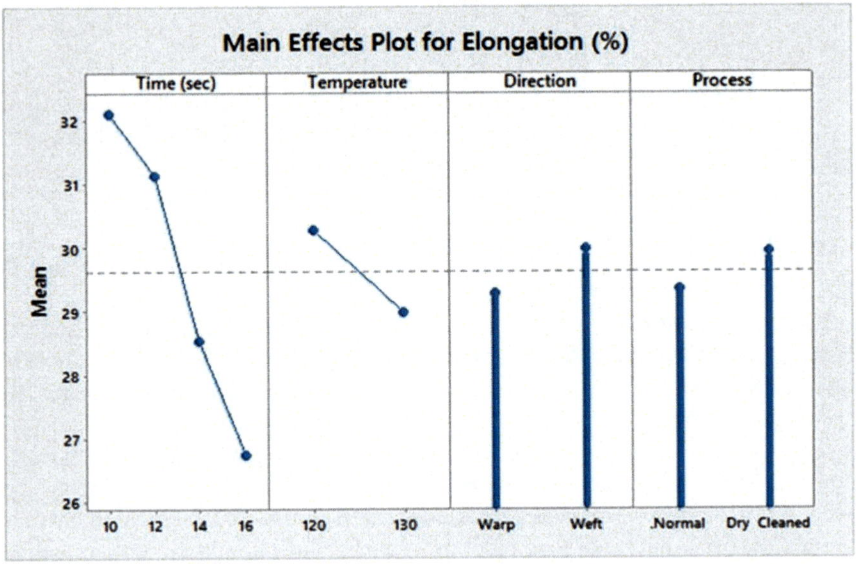

Fig. 4 Elongation (%)

Inherent weft way strength of the fabric is lesser than that of warp way strength. The same trend is observed for fused fabric samples as well. There is no significant impact of dry cleaning on tensile properties of fused samples.

It is clearly visible from Fig. 4 that with the increase in fusing time and temperature there is a reduction in fabric elongation. This reduction in elongation is significant.

This trend in elongation of fabric is observed due to the deposition of polymer/adhesive on fabric after fusing. Interlining adhesive/polymer gets diffused inside the yarn structure, coating the component yarn and fibers of the fabric. This coating is rigid and creates hindrance in natural elongation of component yarn and fibers of the fabric.

Inherent weft way elongation of fabric is higher than that of warp way elongation. The same trend is reflected in the fused sample as well. Weft way elongation is significantly higher than that of warp way elongation. There is no significant effect of dry-cleaning agents on fabric elongation.

Shrinkage

Shrinkage of fused garment patterns is an exceptionally normal and significant issue. This issue could emerge from improper selection of fusing parameters, nonuniform pressure, temperature, and imperfections in interlining [5]. Abundance shrinkage may cause measuring issues as the completed article of clothing will be smaller than it was planned. It additionally prompts the arrangement of puckered creases [6].

Mechanical action during garment manufacturing can cause intemperate shrinkage of fused fabric and results in lesser bond strength. Shrinkage was resolved in two distinct ways. The initial one is hot air shrinkage and the second one is hot water shrinkage.

The mean values of hot air shrinkage and hot water shrinkage of fused fabric in both direction (warp and weft) before and after dry cleaning are given in Tables 9 and 10, respectively. Figures 5 and 6 represent the effect of fusing time, temperature, and dry cleaning on hot air and hot water shrinkage behavior of fabric, respectively.

Table 9 Hot air shrinkage (%)

S. no	Time (s)	Temperature (°C)	Normal		Dry cleaned	
			Warp way	Weft way	Warp way	Weft way
S1	10	120	1.92	1.06	1.89	1.04
S2	10	130	1.82	1.00	1.79	0.98
S3	12	120	1.79	0.95	1.76	0.93
S4	12	130	1.66	0.93	1.63	0.91
S5	14	120	1.16	0.88	1.14	0.86
S6	14	130	1.45	0.80	1.42	0.79
S7	16	120	1.38	0.74	1.36	0.73
S8	16	130	1.18	0.65	1.16	0.64

Table 10 Hot water shrinkage (%)

S. no	Time (s)	Temperature (°C)	Normal		Dry cleaned	
			Warp way	Weft way	Warp way	Weft way
S1	10	120	2.96	1.67	2.93	1.65
S2	10	130	2.23	1.34	2.21	1.33
S3	12	120	2.78	1.49	2.75	1.47
S4	12	130	2.18	1.24	2.16	1.23
S5	14	120	1.80	1.37	1.78	1.35
S6	14	130	1.15	1.11	1.14	1.10
S7	16	120	1.15	1.15	1.14	1.14
S8	16	130	1.03	0.98	1.02	0.97

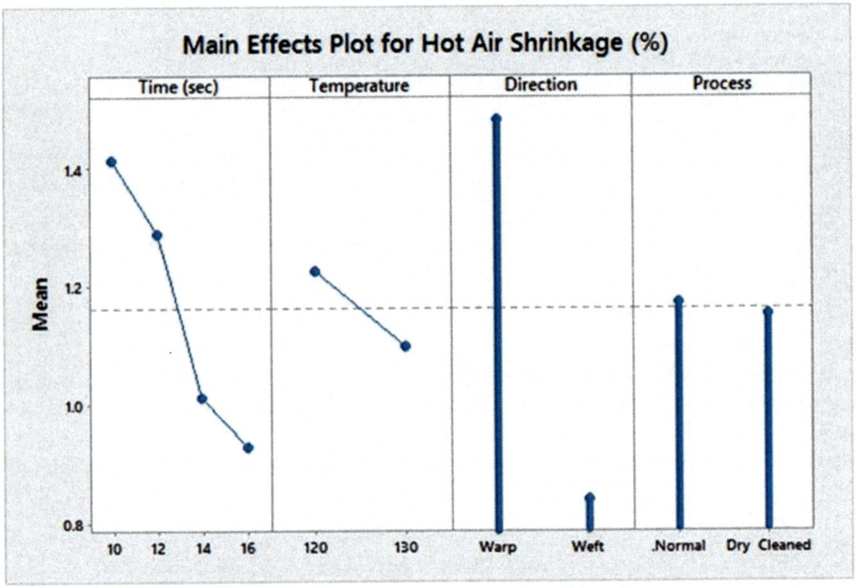

Fig. 5 Hot air shrinkage (%)

Hot water shrinkage test of all the specimens was calculated by immersing the samples in boiling water for 15 min and determining the distance between the gauge marks before and after the test.

From Figs. 5 and 6, all the fabric specimens exhibit similar shrinkage behavior for both hot air shrinkage and hot water shrinkage tests. However, irrespective of all the factors boiling water shrinkage is more than that of hot air shrinkage. Since fabric specimen in water undergoes more relaxation than that of in the air.

In both cases, with an increase in fusing time and fusing temperature, there is a reduction in shrinkage significantly. This trend is may be due to shrinkage in fusible

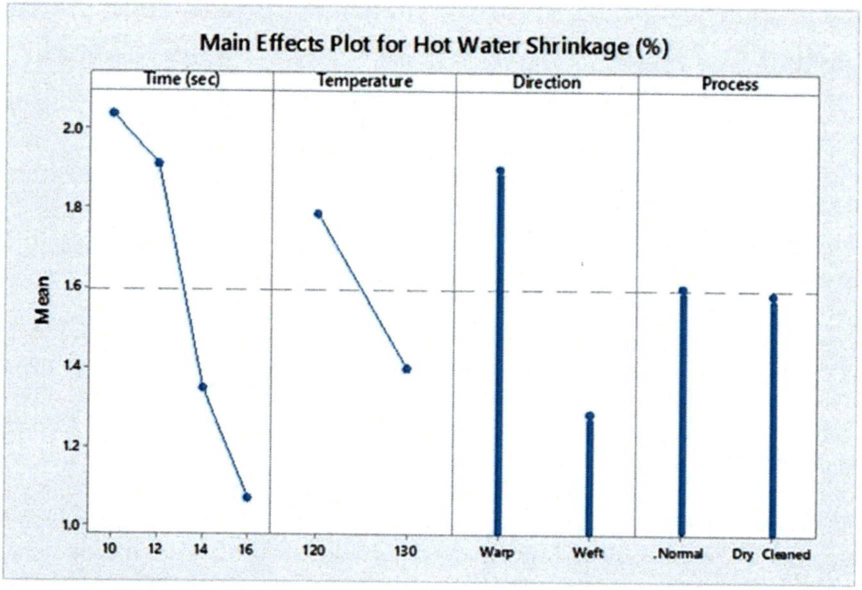

Fig. 6 Hot water shrinkage (%)

interlining during the fusing process. Interlining fabrics are coated with polyamide resin/polymer whose glass transition temperature is 47–60 [6], which is well below the fusing temperature. Hence there is a maximum chance for shrinkage during fusing.

From the results, it has been observed that there seems to be no significant shrinkage on dry cleaning. With the increase in fusing time and temperature, there is a significant reduction in fabric shrinkage. This is because of high residual shrinkage of fabric during fusing. Since already shrunk portion is being subjected to shrinkage, fabrics show reduced shrinkage after fusing.

Irrespective of other factors warp way shrinkage is more than that of weft way shrinkage. Since warp yarns are under much strain due to interlacement than the weft yarns. Hence when a fabric can shrink, the warp yarn shrinks more than that of the weft.

3.2 *Response Surface Optimization*

After studying the effect of fusing time and temperature on the properties of fused textiles, it is essential to optimize the fusing conditions. Keeping this in view, response surface design has been formed for available set of variables with the goal to increase bond strength, formability, i.e., bending modulus, tensile strength and to minimize the shrinkage and elongation. Table 11 exhibits the response surface design for opti-

Table 11 Response surface design

Response	Goal	Lower	Target	Upper	Weight	Importance
Elongation	Minimum		26.040	33.8715	1	1
Tensile strength	Maximum	67.9630	124.800		1	1
Hot water shrinkage	Minimum		0.969	2.9600	1	1
Hot air shrinkage	Minimum		0.540	1.9200	1	1
Bond strength	Maximum	13.4710	21.760		1	1
Bending length	Maximum	6.7535	8.000		1	1

mizing fixing conditions. Table 12 is the solution to the response surface design. From Table 12, we can clearly get the idea about optimum fusing conditions. If the selected interlining is fused of shirting fabric with 130 °C temperature for 16 s one can achieve the desired goals.

Composite desirability has a range from zero to one. If composite desirability is near to one then it will be the ideal case and model achieves the set goals for all the properties. The composite desirability (*D*) of the model is 0.92, which indicates the fusing conditions obtained seem to achieve the desired goal for all the properties as a whole.

4 Conclusion

1. Of all the aspects of performance investigated, it is evident that fusible interlinings fused at with 16 s of fusing time performed well in the areas of bending, bond strength, tensile strength, and shrinkage both in warp and weft direction.
2. Fusing time and fusing temperature is directly proportional to the rate of melting and flow of melt of polymer/adhesive/resin used on interlining.
3. Longer fusing time and higher fusing temperature significantly increase bending length, bending modulus, bond strength and tensile strength.
4. As far as the fusing process is concerned, bond strength shows isotropic behavior.
5. Dry cleaning has a negative influence on several characteristics of fused fabric indicating the unsuitability of either solvent or interlining polymer.
6. To avoid shrinkage in either direction after fusing, fusing temperature must be kept lower than glass transition temperature interlining fabric material as well as shell fabric.
7. Findings of this investigation will help garment manufacturers in selecting the optimum fusing conditions so that they can maintain the cost and quality of the product at the same time.

Table 12 Solution of response surface optimization

Time (s)	Temperature (°C)	Process	Elongation (%) fit	Tensile strength (kgf) fit	Hot Water shrinkage (%) Fit	Hot Air shrinkage (%) Fit	Bond strength (Kgf) Fit	Bending length (cm) fit	Composite desirability (D)
16	130	Regular	26.24	117.91	0.93	0.84	21.51	7.91	0.92

References

1. Amar Z, Al-Gamal G (2015) Effect of different types and orientations of fusible interlinings on men striped shirt cuffs. J Am Sci 1111:66–72
2. Kim M-O, Park M-J (2013) Effects of fusible interlinings on mechanical properties and hand of cotton fabrics for jackets. Text Sci Eng 50:351–358. https://doi.org/10.12772/tse.2013.50.351
3. Kim M-O, Park M-J (2017) Changes in mechanical properties of fusible interlinings on interlock fabrics for T-shirts. Res J Costume Cult 25:448–457. https://doi.org/10.7741/rjcc.2017.25.4.448
4. Fan J, Ng YN (2001) Objective evaluation of the hand of nonwoven fusible interlining. Text Res J 71:661–666. https://doi.org/10.1177/004051750107100802
5. Fan J, Leeuwner W, Hunter L (1997) Compatibility of outer and fusible interlining fabrics in tailored garments part I: desirable range of mechanical properties of fused composites. Text Res J. https://doi.org/10.1177/004051759706700210
6. Glass Transition, https://en.wikipedia.org/wiki/Glass_transition

Doodling: Introducing Paper Art into Textiles

Rashi Garg and Anila

1 Introduction

Art is the expression or application of human creative skills and imagination typically in a visual form and work to be appreciated primarily for their beauty or emotional power [1]. Art presides in every human irrespective of his knowledge about it. The success of a work is determined considerably by its capacity to redefine an idea to give it altogether different dimensions. Art is believed to be consisting of number of features, one of them being imagination. Other trait can be identified as uniqueness or integration. Another striking quality in art form is expressionism not be confused with realism. Expressionism is frenzied, intense and fantastic. Doodle art is the fun way of expressing our inner self. A doodle work usually portrays feelings of the Artist. The more the work is made with feeling, the more exciting work is produced. Doodling is an action to scribble absent-mindedly. Everyone has hidden artist in themselves which is awakened by some means or inspiration or sometimes just by chance. It is enriching exercise. Doodling opens doorways into imagination: there is no such thing as mindless doodles. Doodling art can create wonderful results collaborating with other Arts.

Art and fashion have been linked from years ago. But both disciplines interact and are recognized as interpreters of their time, normally it is the fashion designers who transfers the world of paintings to their creations. From the influence of one art, other arts come up and lead to the new art forms. Therefore with the collaboration of the finest techniques of art and textile printing to produce creations never seen before has opened up amongst many textile art practitioners from different background. Since

R. Garg (✉)
UIFT Panjab University, Chandigarh, India
e-mail: rashigarg99@gmail.com

Anila
Government Home Science College, Chandigarh, India

V. Midha and A. Mukhopadhyay (eds.), *Recent Trends in Traditional and Technical Textiles*, https://doi.org/10.1007/978-981-15-9995-8_15

art has a major role to play in almost all spheres, therefore, the possibilities of creating new areas captivating the interests of art lovers and fashion practitioners in fashion is unfolded through the present study by collaborating the doodling art with textile decoration technique colour blending. And this research strives to work towards spreading awareness regarding doodling as an art. The physical act of producing a doodle necessarily engages the mind, opening doorways into imagination, the intellect, and the hidden centers of insight [2, 3].

The study "Doodling: Introducing paper art into textiles" involves development of doodling designs inspired from students scribbled designs with colour blending for living area range. It would serve as a guideline for new designers to develop innovative creations and introduce something new to the world for manufactures as well as the consumers of textile products. In today's hyper-digital environments, there seems to be a near-universal sensation of fractured attention for many of us, where doodling elevates focus and concentration. Understanding of doodling art is moderate so the awareness of concept and practice of doodling art can be spread among the youth [4, 5].

1.1 Objectives

- To study the Doodling Art and Colour Blending.
- To collect scribbled design from student notebooks.
- Design development inspired from scribbled design and documentation of both.
- To create products for living area range using doodling art and colour blending.
- To find out marketability and consumer acceptability of the created range.
- To spread awareness regarding doodling as stress buster art.

2 Methodology

To achieve the aims of the study, the methodology was divided into the following sections.

- Study of doodling art and colour blending.
- Collection of students scribbled designs.
- Selection of product range.
- Sourcing of suitable fabric.
- Evaluation of colours for background.
- Pretesting of samples with colour blending and doodling.
- Design development taking inspiration from students scribbled designs and evaluation of designs.
- Documentation of both scribbled and doodled designs.
- Creation of products for living area range using doodling art and colour blending.

- Costing of products.
- Marketability and consumer acceptability of the products.
- Awareness regarding doodling art.

3 Results and Discussions

3.1 Study of Doodling and Color Blending

To study the history, benefits, techniques and other characteristics of doodling art and colour blending data was collected from books, internet, and art gallery.

3.2 Collection of Students Scribbled Design

Scribbled designs from student's notebook were collected randomly from hostel and college students of Govt. Home Science College, Chandigarh (Fig. 1).

3.3 Selection of Product Range

For the selection of product range a market survey was conducted to explore the various ranges by visiting various home décor shops of Chandigarh. Shopkeepers were asked to fill the questionnaire regarding end use of the product and to make it affordable with higher market demand (Fig. 2).

3.4 Sourcing of Suitable Fabrics, Colors for Background

Sourcing of suitable fabric was done in which different fabrics were evaluated by panel of judges. For selection of background colour for blending, six different colour themes were created and evaluated by judges.

It is clear from Table 1 that sample no. 1 i.e. Glazed Cotton got maximum marks (284) and was at 1st rank. Sample no. 8 i.e. Cotton was at 2nd position with 240 marks. Sample no. 3 i.e. Cotton Linen was at 3rd position with 223 marks. Sample no. 2 i.e. Khadi Cotton was at 4th position with 214 marks. Sample no. 5 i.e. Crepe was at 5th position with 200 marks. Sample no. 4 i.e. Muslin was at 6th position with 134 marks. Sample no. 6 i.e. Matty was at 7th position with 113 marks. Sample no. 7 i.e. Linen was at 8th position with 78 marks. Thus sample no. 1 i.e. Glazed cotton was selected for living area range due to its good absorbency, soft touch and blending effect and doodling was good on glazed cotton when compare to others.

Fig. 1 Collection of students scribbled design

Fig. 2 Distribution of respondents on the basis of suitability of doodling art in different ranges

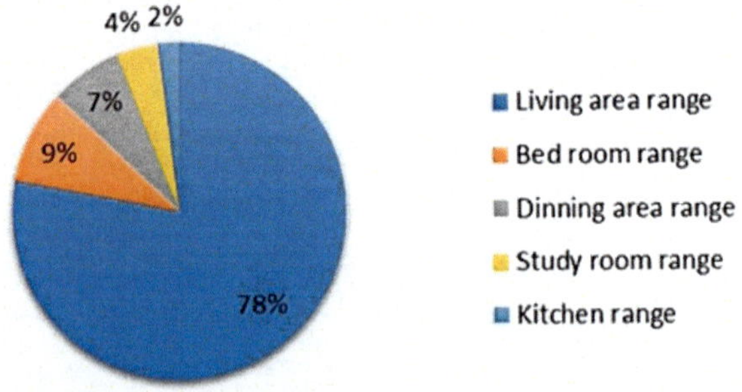

Table 1 Selection of most preferred fabric for living area range

SI. No	Sample name	Marks	Rank
1	Glazed cotton	284	1
2	Khadi cotton	214	4
3	Cotton linen	223	3
4	Muslin	134	6
5	Crepe	200	5
6	Matty	113	7
7	Linen	78	8
8	Cotton	240	2

It is clear from Table 2 that sample no. 5 was at 1st position with score 206. Sample no. 3 was at 2nd position with score 204. Sample no. 2 was at 3rd position with score 184. Sample no.4 was at 4th position with score 181. Sample no. 6 was at 5th position with score 144. Sample no. 1 was at 6th position with score 123. Majority of respondents referred Sample no. 5 and 3 for living area range due to their soft and soothing effect.

3.5 *Design Development Taking Inspiration from Students Scribbled Designs and Evaluation*

A total 10 range designs were designed inspired from students scribbled designs. For selection of range designs, range designs were evaluated by panel of judges. Documentation of both scribbled and developed designs was done (Fig. 3; Table 3).

3.6 *Creation of Products*

Selected ranges were constructed into final products (Fig. 4).

3.7 *Marketability and Consumer Acceptability*

Market and consumer acceptability was done and responses from both for living area range were found appreciable, very innovative, unique and exclusive (Fig. 5).

The above figure reveals that 78% of the shopkeepers wanted to place order. They were ready to buy the products at the coated price whereas 22% of the shopkeepers like the products but wanted to place order later on.

The above figure reveals that:

Table 2 Selection of most preferred colour theme for background for living area range

SI. No	Colour theme	Marks	Rank
1		123	6
2		184	3
3		204	2
4		181	4
5		206	1
6		144	5

Majority of shop keepers i.e. 100% of them appreciated the overall appearance of wall panels. They found that the designs were very unique, innovative and exclusive (Fig. 6).

Majority of shop keepers i.e. 100% of them appreciated the overall appearance of cushion covers. They found that the designs were very unique, innovative and exclusive.

Majority of shop keepers i.e. 93% of them appreciated the overall appearance of lampshades very good and rest of i.e. 7% appreciated the overall appearance of lampshades good. They found that the designs were very unique, innovative and exclusive.

Majority of shop keepers i.e. 100% of them appreciated the overall appearance of tea coasters. They found that the designs were very unique, innovative and exclusive.

Fig. 3 Design development taking inspiration from students scribbled designs and evaluation

Fig. 3 (continued)

Table 3 Evaluation of designs for living area range

SI. No	Range	Marks	Rank
1	RANGE-1	239	6
2	RANGE-2	273	2
3	RANGE-3	320	1
4	RANGE-4	247	4
5	RANGE-5	245	5
6	RANGE-6	255	3
7	RANGE-7	200	7
8	RANGE-8	111	10
9	RANGE-9	133	9
10	RANGE-10	153	8

Fig. 4 Creation of products

Fig. 5 Distribution of the respondents on the basis of their willingness to buy/place order

3.8 *Awareness Regarding Doodling Art*

The objective of dissemination of the gathered data was achieved by conducting workshops for students on doodling art. The workshop conducted received positive responses and ample interest was shown by students for attending more such programs.

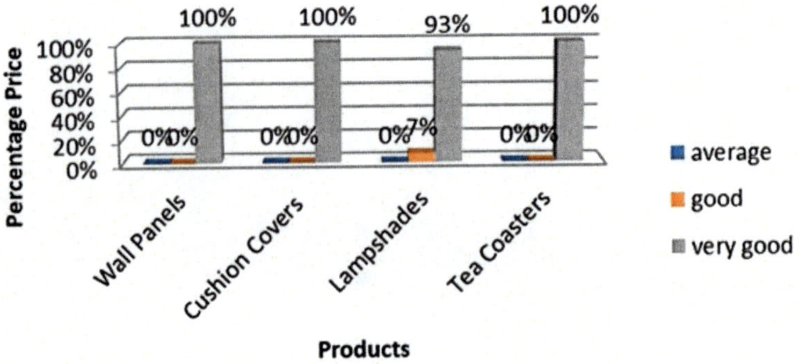

Fig. 6 Distribution of respondents on the basis of their liking of the overall appearance of the products

4 Conclusion

The study "Doodling: Introducing paper art into textiles" involves development of doodling designs inspired from students scribbled designs with colour blending for living area range. Major finding of this study was developed designs inspired from students scribbled designs artistically combined with colour blending for living area range. The awareness of concept and practice of doodling art can be spread further among the students of fine arts and to the society as the present understanding was moderate. The created doodling art techniques and products encouraged the viewers to ask questions regarding the skill involved and conception delivered. The techniques were demonstrated to students in workshops who immensely enjoyed the ideas and showed interest in attending more such workshops.

References

1. Definition of Art (nd) Retrieved 4 May 2018, from Wikipedia website https://en.wikipedia.org/wiki/Art
2. Dickie G (1984) The art circle. Haven Publications, New York
3. Read H (1967) Art and alienation. Thames and Hudson, London
4. Doodle (nd) Retrived 18 May 2018, from Wikipedia website https://en.wikipedia.org/wiki/Doodle
5. How to Doodle (nd) Retrieved 8 May 2018, from Wikihow website https://www.wikihow.com/Doodle

Manufacturing Technologies and Scope of Advanced Fibres

Amal Chowdhury and S. Dhamija

1 Introduction

The evolution of fibre developments has gone through the phases of conventional, highly functional and advanced/high-performance fibres. The initial motivation for the development of high-performance fibres came from aerospace industry seeking fibres for light but stiff, strong and tough composite structural parts [1]. Later, constant pursuit for critical applications has enabled further evolution of advanced fibres with a wide spectrum. These advanced fibres are largely used in technical textiles––reaching to different application domains of vehicles, construction, agriculture, sports, health care, defense and security, electronics, power, environmental technologies, filter materials, protective clothing, reinforcement for tyres, rubber goods, composites and many more—thus providing new opportunities to the global textile industry.

For the production of advanced fibres, manufacturers need a comprehensive understanding of the fibre manufacturing technologies in the backdrop of the available fibre-forming polymers. Besides the introduction of new technologies such as gel- and microfiber spinning, there have been developments in melt- wet- and dry spinning. This paper is expected to provide an exposure to the manufacturing technologies of some revolutionary fibres.

A. Chowdhury (✉) · S. Dhamija
Department of Textile Technology, The Technological Institute of Textile & Sciences, Bhiwani, Haryana, India
e-mail: amal_chowdhury90@rediffmail.com

V. Midha and A. Mukhopadhyay (eds.), *Recent Trends in Traditional and Technical Textiles*, https://doi.org/10.1007/978-981-15-9995-8_16

2 Aramid Fibres

These fibres are generally based on aromatic structures with well-oriented rigid chains which bestow excellent mechanical properties and thermal resistance. Aramids decompose before or during melting. Hence, they are spun from the solution. Over the period, a number of aramid fibres like Nomex, Kevlar, etc. have been developed.

2.1 Nomex

Nomex is synthesized by condensation process in which m-Phenylenediamine and Dichloride of m-isophthalic acid react to yield Poly (m-phenylene isophthalamide) or Nomex polymer. The polymer is dissolved in dimethylformamide (DMF) to make a dope solution of 20% polymer. 4.5% of Lithium Chloride (LiCl) is also added to improve its solvency by reducing the viscosity of the solution. The solution is then dry spun using hot air at 200–210 °C. Otherwise, it can also be wet spun by coagulating in water. The as-spun fibre is extracted in cold water with a bid to remove LiCl and lastly drawn. Drawing is carried out with a draw ratio of about 5.5 using steam.

Nomex has excellent thermal resistance [2], but poor mechanical properties. Poor mechanical properties can be attributed to the presence of meta-oriented aromatic rings, which disallow compact packing of molecules. Further, Nomex fibres show good chemical resistance and can also hinder radiation.

As far as applications of Nomex are concerned, it finds its applications in protective clothing in hostile environments having heat, chemicals, and radiation, spacesuit [3, 4], furnishings in public places, industrial fibres. It is also used to make hollow fibres for desalination by reverse osmosis.

Other variants of Nomex, viz., Nomex Delta FF and Nomex Delta Micro have also been developed for improved filtration efficiency.

2.2 Kevlar

The inherent weakness of Nomex fibre is its relatively poor mechanical property. This very fact became a motivation for the discovery of Kevlar fibre. Replacement of meta-oriented aromatic rings with para-oriented rings, perhaps, causes better intermolecular registration of amide groups and make it more crystallizable. This very thought opened a new space for the invention of Kevlar fibre.

Kevlar is synthesized by the condensation of terephthaloyl chloride (TCL) and p-phenylenediamine (PPD) in a mixture of hexamethyl phosphoramide (HMPD) and n-methyl pyrrolidone (NMP) solvents. This synthesis process produces poly (p-phenylene terephthalamide) fibre (PPTA). To avoid degradation and to minimize the side reactions, the polycondensation should be carried out at 10–20 °C. PPTA is dissolved in sulphuric acid. PPTA forms liquid crystal solution under suitable conditions of concentration, temperature, solvent and molecular weight. The liquid crystal form is akin to nematic-like structure. Maximum anisotropy occurs at a polymer concentration of 20% in 100% $H2SO_4$. At this stage, the 100% liquid crystalline phase is achieved. Liquid crystals are characterized by flowable, optically anisotropic structure. They essentially have insufficient energy for individual molecules. This restricts the freedom of rotation of the molecules and thereby assists with the formation of aggregated-parallel-arranged molecules [5].

The process of spinning Kevlar is dry-jet wet-spinning. The polymer is extruded at about 100 °C maintaining an air gap of 1.0 cm or 0.5 cm (Fig. 1). Subsequently, it is coagulated into cold water at the temperatures in the range of 0–5 °C. These conditions lead to bring about good molecular orientation along the spin-line.

During coagulation, the solution undergoes a relaxation effect resulting in a smectic-like structure, giving rise to skin-core super-molecular structure of Kevlar fibre. The properties of the fibre can be varied by introducing variation in the spin stretch factor (SSF). With the increase of SSF, tenacity and modulus both increase. However, these improvements are at the expense of extension to break. SSF is usually maintained less than 10 in a bid to control breakages of filaments. Kevlar, in general, is spun at a speed of around 50 m/min. After being spun, filaments are washed and dried. Subsequently, they are given heat treatment under tension in the range of 450–550 °C for a few seconds. This step helps in the improvement of the orientation of molecular chains to enhance fibre properties. It is important to mention here that Kevlar could also be wet spun. However, mechanical properties of wet spun Kevlar are inferior to those obtained from dry-jet wet-spinning.

Kevlar is accepted as high-performance fibre because of its outstanding modulus, strength, toughness, and temperature resistance. It is lighter in weight, durable and cheaper than steel.

Kevlar finds its best applications in belting in radial tyres and as cord in heavy-duty track tyres and aircraft tyres. These are also used for composite applications in

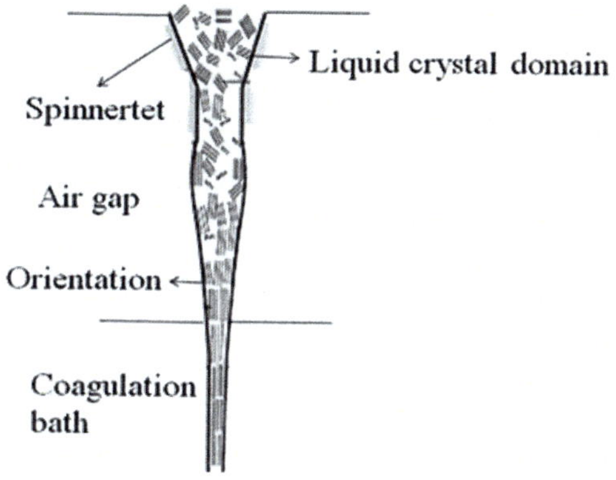

Fig. 1 Liquid crystalline domains in dry-jet wet spinning

civilian and military aircraft, helicopter parts, protective apparel, ropes and cables, industrial fabrics and pressure vessels.

Different grades of Kevlar, namely, Kevlar 29, Kevlar 49, Kevlar 149, Kevlar 981, etc. have been developed to suit various high-performance applications.

3 PBO

PBO [Poly (p-phenylene benzobisoxazole)] is called 'ordered polymer' because of its ability to form highly ordered structures in the solid state. The persistence length of PBO is around 640 Å, much higher than that of Kevlar (200 Å), which confers high rigidity to the former.

PBO is synthesized by the condensation of 4,6-diamino-1,3-benzenediol dihydrochloride (DABDO) and terephthalic acid (TA). The reaction is carried out in polyphosphoric acid (PPA) solvent, yielding PBO which has about 200–400 repeat units per chain, which translates into the molecular weight of 50,000–100,000. The dope exhibits liquid crystalline phase which is a function of various parameters such as concentration of polymer, solvent and temperature and is extruded through spinneret using dry-jet wet-spinning technique. Polymer is coagulated in water. Subsequently, filaments are washed and dried. At the next step, fibres are heat treated under tension in an inert atmosphere. The temperature during heat treatment is maintained in the range of 500–700 °C with a residence time of a few seconds to several minutes. Heat treatment causes stabilization of the structure, which may be attributed to post-crystallization and stress relaxation in the fibre.

PBO is known for its high mechanical properties [6] and thermal resistance. However, it possesses poor compressive strength. It is used as reinforcement in composites, multilayered circuit boards, athletic equipment, marine applications and cables. It is also used in fire protection fabrics. Fabrics made from these fibres are used for ballistic protection. As PBO has poor compressive strength, its uses are limited to those applications where axial compression loading does not occur.

4 Vectran

Vectran is made from wholly aromatic polyester. Normal polyester fibre is made from the condensation process of TPA and Glycol, which produces aliphatic-aromatic polymer. Aromatic terephthalic acid residue confers rigidity to the fibre while glycol residue is responsible for its flexibility. It has not been possible to make high-performance fibre of PET, even with high molecular weight. However, wholly aromatic polyesters overcome this shortcoming.

Vectran fibre is made from Vectra liquid crystal polymer [5, 7]. This polymer is made by the acetylation polymerization of p-hydroxybenzoic acid and 6-hydroxy-2-napthoic, which upon melting, attain liquid crystalline phase over a certain range of temperature. The spinning process is akin to melt spinning. The thermotropic liquid crystalline spinning melt is extruded at 280–350 °C through a spinneret. Spinning speed is about several thousand metres per minute. The as-spun fibres are given heat treatment at temperatures of 250–300 °C for several hours. During heat treatment, solid-sate polymerization occurs, which increases molecular weight (M_n) by three times, along with an increase in crystallinity by almost 20%. All these changes enhance fibre strength.

The fibres are largely used for ropes and cables because of their high strength, good abrasion resistance and negligible creep. As Vectran fibres have the excellent vibration-damping capability, they are used as bowstrings, bicycle frames and sailcloth.

5 Spectra

High-performance polyethylene fibres are commercially produced under the trade name 'Spectra' by Honeywell in the USA. The only flexible chain which has been commercialized as high-performance fibre is polyethylene. The flexibility of the backbone of chain may be accounted for by two reasons. First, there is a possibility of rotation around C–C bonds in its structure. Second, the presence of light hydrogen as the only other element confers flexibility.

Spectra is synthesized by polymerizing ethylene. Under specific conditions of polymerization, it yields a polymer of high molecular weight (approximate M_w 10^6). However, high molecular weight of polymer always associates high melt viscosity that hinders spinning of fibre from the melt. Furthermore, a flexible polymeric chain of high molecular weight leads to a very high degree of entanglement of the molecular chains—which restricts the drawing of a melt-processed UHMW-PE to a very limited extent.

However, the solution comes with the gelation/crystallization method, a well-known and powerful technique [8], to prepare ultra-high-molecular-weight polyethylene (UHMWPE) fibre. This innovative technique can produce fibres with high modulus and high strength as it offers the advantage for dry gel form of being drawn with a high draw ratio. The polymer is in a 'gel' state—only partially liquid—which keeps the polymer chains somewhat bound together. These bonds give rise to strong inter-chain forces in the fibre, which increase its tensile strength. While manufacturing through gel spinning, the polymer is dissolved in decalin or paraffin oil to make 5% solution at 130–140 °C. The solution is pressurized through the spinneret into a small air gap and then enters a water bath (Fig. 2) at room temperature to form a gel fibre.

The fibre is hot drawn with a draw ratio of 30–100 at temperatures of 130–140 °C.

Spectra has lightweight, high strength [9], high modulus, high toughness and high chemical resistance. However, the melting point is low as the chains are flexible. Thus, thermal resistance is a limitation of this type of fibre.

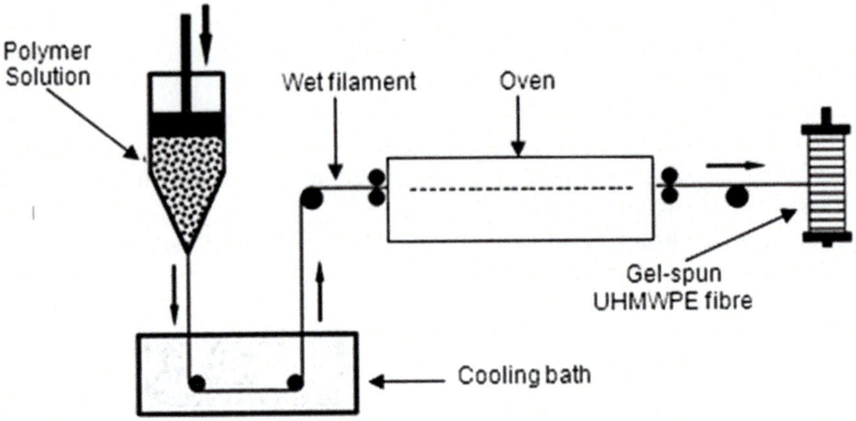

Fig. 2 Gel spinning process

One can find the applications of Spectra in marine ropes, cables, sailcloth, concrete reinforcement, fish netting, sports equipment and medical implants. The high modulus of Spectra permits its use in ballistic protection, cut-resistant gloves. Apart from these, Spectra is also used in space research as layers of polyethylene are supposed to protect astronauts. Polyethylene has a lot of hydrogen, which is a good radiation blocker, thus making it a promising material for spacesuit [10].

Further developments in manufacturing specialized UHMWPE has also been carried out by incorporating multi-walled nanotube [8]. Well-blended UHMWPE-MWNT composites prepared by gelation/crystallization provide characteristics of high modulus along with high electric conductivity.

6 Carbon Fibre

Unlike, other polymeric fibres, carbon fibres have planar graphite structure. The arrangement of the layer planes in the cross section of the fibre is important since it affects the transverse and shear properties of the fibre.

Among various precursors, evaluated for manufacturing carbon fibres, only three precursors, viz., Viscose rayon, Pitch and Polyacrylonitrile (PAN) are commercially successful. Out of these, viscose rayon (carbon yield ~30%) is losing its importance as it produces an inferior quality of carbon fibre. On the other hand, pitch-based precursors produce carbon fibres with a higher yield (80%) and relatively higher strength.

PAN-based carbon fibre is the best when mechanical properties are considered. The production of Viscose rayon and PAN-based carbon fibres involves three stages, viz., oxidative stabilization, carbonization, and graphitization. The temperature may be varied during each stage to produce the type of fibre required.

When it comes to pitch-based carbon fibre, the production route is somewhat different from PAN-based carbon fibre. There are two types of pitch: isotropic and anisotropic. Isotropic pitch goes through centrifugal spinning or melt blowing technique whereas, anisotropic pitch undergoes melt spinning giving liquid crystalline phase to produce carbon fibre. Mesophase pitch fibres can further be heat-treated to produce very high modulus carbon fibres.

Carbon fibre has high specific strength and stiffness. It also exhibits high temperature resistance, chemical and biological inertness, better electrical conductance, good vibration damping ability and fatigue resistance. Two main sectors of carbon fibre applications are the high technology sector that includes aerospace and nuclear engineering. Other applications include bearings, gears, cams, fan blades, automobile bodies and sports equipment.

Carbon fibre has a drawback that they are having an inert surface. So, they do not allow matrix material to make bonding with it. Surface treatment of fibres is one of the suggested methods to improve adhesion between the two [11]. There are various techniques for surface treatment, e.g. gaseous oxidation, liquid-phase oxidation, whiskerization or polymer grafting.

With the development of technology, it has been possible to produce carbon fibres with diameters in the range of 4–50 nm and lengths of several micrometers, using arc and laser ablation process and chemical vapour deposition process. The fibres have well-defined multiple or single wall. Correspondingly, these are known as multi-wall carbon nanotubes (MWCNT) and single-wall carbon nanotubes (SWCNT). These fibres have contributed a lot to high-performance applications.

7 Glass Fibre

Glass fibres come under the category of inorganic fibres. It is relatively inexpensive amongst other inorganic fibres. These fibres have three-dimensional structures in contrast to the uni-dimensional polymeric fibres and two-dimensional carbon fibres.

Glass fibres are produced by the melt-spinning route. It is a special processing technique which is different from those used for other polymeric fibres described earlier.

There are four steps involved in the manufacture of glass fibres. First, dry mixing of ingredients (SiO_2, Al_2O_3, CaO, etc.) is carried out. Second, melting is done in a refractory furnace at 1370 °C. Third, extrusion follows through orifices in an electrically heated platinum bushing and rapid attenuation occurs under gravity [5], accompanied by cooling. Fourth, after solidification, 0.5–2% by weight of size, a binder is applied to the strand. Glass fibres are produced in various forms, namely, strands, roving, chopped strands (3–2 mm), milled fibres (0.8–3.2 mm).

These fibres have moderate density, high strength and corrosion resistance. Applications of glass fibres can be divided into four basic categories: (a) insulations, (b) filtration media, (c) reinforcements and (d) optical fibres. One of the principle high-performance applications for these fibres is in their use as reinforcements for composite materials.

Various types of glass fibres have been developed such as soda-lime-, D-, E- and S- glass fibre, etc. It is worth pointing out that unlike soda-lime glass, the total alkali content in E- and S-glass fibre is kept below 2% to ensure good corrosion resistance and a high electrical surface resistivity. Among these, E-glass fibres are used in fire-resistant textiles, which can be coloured using dyeable sizing.

8 Conclusion

To meet the needs of the consumer, the fibre producer must have a good understanding of how the technologies control the fibre properties. Based on this understanding, the fibre can be engineered by a suitable choice of different variants. To add, far-reaching developments can take place in the growth of high-performance fibres through technological innovations. As the researchers are searching new classes of polymeric materials with unique applications, it is likely that new advanced fibres will be engineered in near future in positive and productive ways. Because of the phenomenal growth of high-performance applications, the future of advanced fibres appears to be very promising indeed. However, an in-depth research is still needed for their development and is poised for a revolutionary change.

References

1. Gupta VB, Kothari VK (1997) Manufactured fibre technology, 1st edn. Chapman & Hall, London, pp 514–557
2. Technical textiles at the forefront of innovation. Available from www.plastics-themag.com/applications/sports-textiles-polymers-in-the-lead. Accessed 07 Aug 2019
3. Spacesuit. Available from https://www.madehow.com/Volume-5/Spacesuit.html. Accessed 05 Sept 2019
4. Pavan MVR, Macharla K, Hayavadana J (2009) Scope of technical textiles in space & aerospace. Available from https://www.indiantextilejournal.com/articles/FAdetails.asp?id=1958. Accessed 20 Aug 2019
5. Hearle JWS (2007) High-performance fibres. The Textile Institute, CRC Press, Woodhead Publishing Limited, England, pp 1–235
6. Moving high performance fibre forward. Available from https://www.fiber-line.com/en/fibers/zylon-pbo. Accessed 05 Sept 2019
7. Rameshkumar C (2008) Revolutionary high performance fibres. Available from https://www.indiantextilejournal.com/articles/FAdetails.asp?id=792. Accessed 10 July 2019
8. Matsuo M, Yamanaka A, Nakano Y (2009) Gelation mechanism of ultra-high-molecular-weight polyethylene (UHMWPE) chains in dispersion solutions containing multiwall carbon nanotubes (MWNTs) analyzed in terms of liquid–liquid phase separation. Pure Appl Chem 81(3):513. https://doi.org/10.1351/PAC-CON-08-06-03
9. Ultra-high-molecular-weight polyethylene. Available from https://en.wikipedia.org/wiki/Ultra-high-molecular-weight_polyethylene. Accessed 02 Aug 2019

10. Future spacesuits could heal themselves. Available from https://www.newscientist.com/article/dn9624-future-spacesuits-could-heal-themselves/. Accessed 09 July 2019
11. Tiwaria S, Bijweb J (2014) Surface treatment of carbon fibers—a review. In: 2nd international conference on innovations in automation and mechatronics engineering, ICIAME 2014, Procedia Technology, vol 14, pp 505–512. Available from www.sciencedirect.com. Accessed 06 Sept 2019

Investigation of Flammability Parameters of Different Types of Fabrics Using Digital Image Processing Technique

Ajit Kumar Pattanayak

1 Introduction

Flammability of any material is the potential of that material to burn or ignite, causing fire or combustion. The severity of the burning of the substance is quantified through fire testing. There are several test methods available for the quantification of the flammability characteristics [1]. Flammability testing method must be the most important testing procedures for the textile industry as it can avoid several lives in the event of fire. It has been observed that most of the deaths are caused due to the accidental burning of upholstery and textiles. Hence, proper and easier flammability standards should be framed to avoid these hazards [2]. Flammability performance can be improved by the fabric manufacturer at the design stage to ensure a safer interior environment. But there is no such standard is available for the assessment of flammability digitally. There is an assortment of clothing flammability and related tests [3, 4]. These test methods attempt to determine how flame-resistant, heat-resistant, or injury-resistant a particular fabric and/or garment is. The flammability test is conducted in three basic formats such as small-scale testing, large scale testing and the instrumented manikin test [5]. Small-scale testing is the backbone of research, as it provides the most cost-efficient method to compare different specimens amongst one another. However, protective clothing small-scale test methods provide only a limited analysis of a fabric's performance in a specific scenario; thus, several different tests may be required to get a general understanding of a fabric's performance [6]. Currently, the large-scale test method for clothing flammability uses a fully dressed manikin to represent a person. This manikin is then subjected to

A. K. Pattanayak (✉)
Department of Textile Technology, The Technological Institute of Textiles & Sciences, Bhiwani, Haryana, India
e-mail: ajitpatnaik@gmail.com

V. Midha and A. Mukhopadhyay (eds.), *Recent Trends in Traditional and Technical Textiles*, https://doi.org/10.1007/978-981-15-9995-8_17

a fire environment and an evaluation of the entire garment assembly is made rather than just a small sample of textiles from the small-scale test methods.

The proposed method to determine the different parameters by using an image processing method is a small-scale testing procedure in which different flammability parameters are determined accurately by analysing the captured video of the burning sample.

2 Materials and Methods

2.1 *Materials*

Apparel grade fabrics are used for this study for the estimation of different flammability parameters. Five samples covering a variety of structures are used for this study. The fabric specification is shown in Table 1.

2.2 *Methodology*

2.2.1 Design of the Digital Vertical Flammability Tester

To measure the flammability of the fabric using image processing technique, an instrument was fabricated based on the basic idea of Vertical Flammability Tester [7]. It is designed in such a way that it can be used for conventional as well as for image processing technique for accurate results. The specification of the specially developed vertical flammability tester is shown in Fig. 1. The video capturing is carried out by using a digital video recorder in MP4 format. The standard dimension of the image is 1920 × 1080 and the depth of the image is 36 bits. The dimension of the sample size used for this study is 330 × 50 mm. The samples are then oven dried for 30 min at 105 °C to remove the moisture present in the fabric. Then, the samples are mounted in the specially developed vertical flammability tester with the help of the clamps and ignited as per the norms [8].

Table 1 Fabric specification used in this project

Sr. no	Sample	Weave	Areal density (gm/m^2)
1	100% cotton	Plain	213
2	100% cotton	Twill (3/2)	205
3	100% PET	Plain	78
4	Cotton + 5% Lycra	Denim	320
5	100% Linen	Plain	202

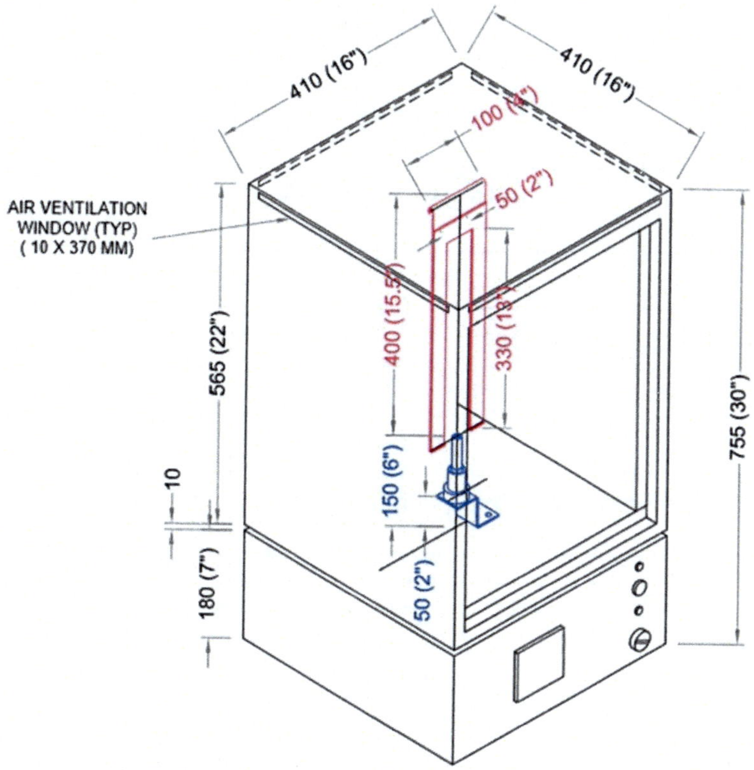

Fig. 1 Schematic of vertical flammability tester (Front view)

2.2.2 Video Processing Algorithm

In this measurement technique, first the video is grabbed by the CCD camera and that video is processed by the software developed in the MATLAB to calculate various flammability parameters. The sequence of video processing operations used for the determination of flammability parameters are given in Fig. 2.

2.2.3 Definition Flammability Parameters

Burning Time: It is the recorded time taken for the self-extinguish of the sample after ignition.

Rate of burning: It is the ability of the flame to spread from the ignition point. This parameter is an important aspect for the risk assessment of any fabric under fire. This parameter can be calculated as per Eq. 1.

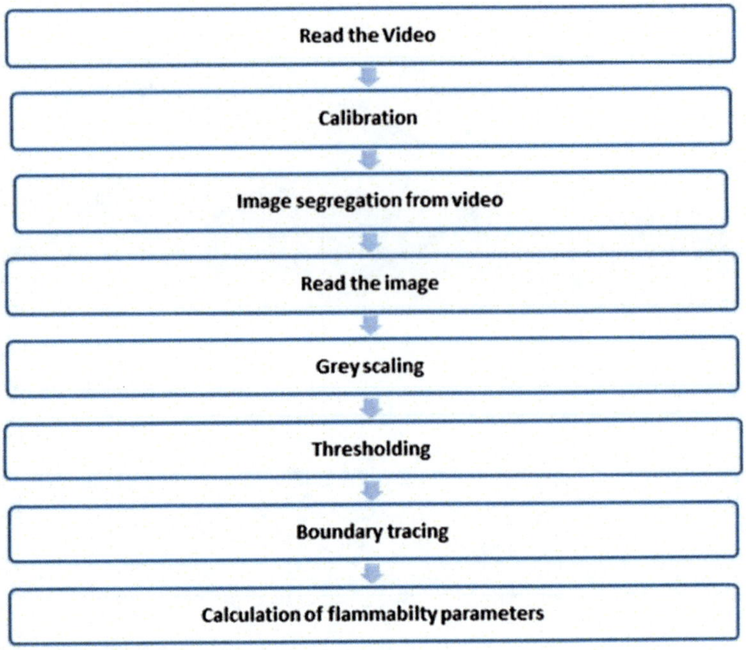

Fig. 2 The flowchart of the working principle of the digital flammability tester

$$\text{Rate of buring} = \frac{\text{Length of fabric burnt}}{\text{Time for complete burning}}\text{mm/sec} \quad (1)$$

Thermal Profile: The most important parameter of the thermal profile is Peak Temperature attended during burning. The other parameters are the afterglow, emissivity, etc.

Char length: This an important indicator of the burning behaviour of the fabric sample. The char produced by the burning fabric sample helps to control the further burning. Hence, this parameter is very helpful to access the fire spreading. This parameter can be calculated using simple measurement manually. But with this newly devised instrument, the image processing software can easily and accurately measure the parameter.

Geometrical Parameters: Burning Area—actual number of pixels in the burning region; Centroid—centre of mass of the region (Cartesian coordinate); Orientation––the angle between the *x*-axis and the major axis of the ellipse that has the same second-moments as the region and Perimeter—distance around the boundary of the region.

3 Results and Discussion

3.1 Estimation of Fabric Flammability Characteristics

The fabric samples were evaluated for various flammability parameters using the newly developed digital image processing based vertical flammability tester and also the conventional vertical flammability tester. The results for flammability parameters measured by the conventional technique and by digital image processing technique are given in Fig. 3.

3.2 Thermal Profile of the Burning Samples

This study also aims to analyse the thermal profile of the samples. This purpose is achieved by capturing the thermal images of the burning samples at different time interval. The maximum temperature is observed at the middle of the burning sample area which attains about 280 °C. This temperature is very dangerous for the human skin in case of fire in the garments. The emissivity represents potential heat that can be transmitted to the burning sample to the next layer, i.e. to skin.

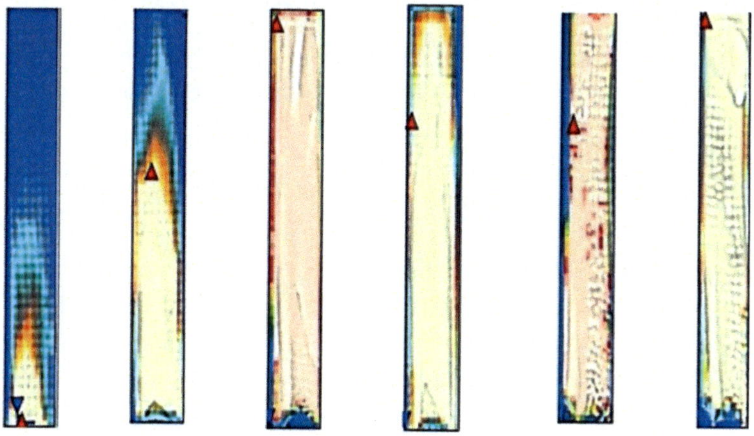

Fig. 3 Thermal image of Sample 2 at different burning time

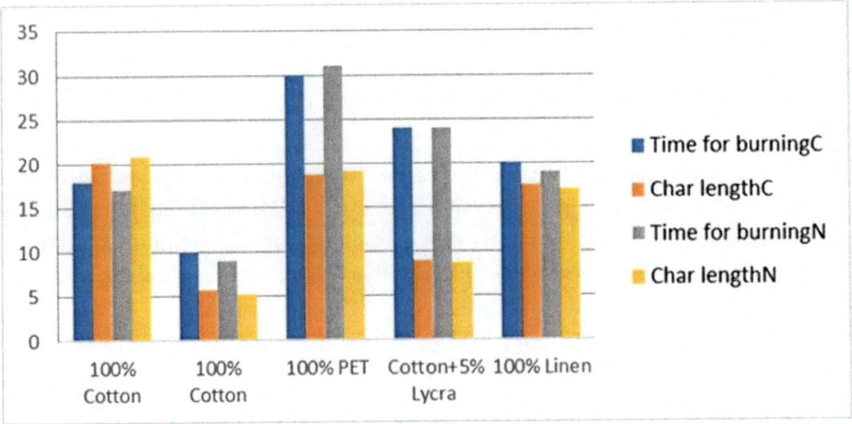

Fig. 4 Comparative assessment of flammability by image processing method and conventional method (C—conventional and N—new)

3.3 Comparative Assessment of Flammability by Image Processing Method and Conventional Method

The flammability assessments of the five samples are estimated by using image processing technique and also with the conventional technique. The results are shown in Fig. 4. It is found that both the techniques have similar burning time and char length. But apart from these two parameters other parameters (char area, centroid of the burning mass) can also be estimated by image processing technique. Hence this method is more useful for the assessment of the complete behaviour of the flammability of textile fabrics.

4 Conclusion

In this study, a trial has been made to characterize the fabric flammability behaviour of different types of fabrics by using image processing technique. This method is also validated by comparing the burning time and char length with the conventional technique. It is observed that both the methods have similar results. The image processing method is also capable to estimate different flammability parameters like char area, centroid of the burned area for a complete understanding of the flammability behaviour of the fabrics. So this method can replace the conventional technique for fabric flammability test.

References

1. Barker S, Tesoro GC, Toong TY, Moussa NA (2003) Textile fabric flammability. MIT Press, Cambridge, MA, pp 201–288
2. Nazare S, Horrocks AR (2008) Flammability testing of fabrics. In: Hu J (ed) Fabric testing. Woodhead Publishing, Cambridge, UK, pp 339–385
3. Nazare S, Kandola BK, Horrocks AR (2002) Use of cone calorimetry to quantify the burning hazard of apparel fabrics. Fire Mater 26:191–200
4. Norton MJT, Kandolph SJ, Johnson RF, Jordon KA (1985) Design, construction, and use of Minnesota woman, a thermally instrumented mannequin. Text Res J 55(1):5–12
5. Horrocks AR (2001) Textiles. In: Horrocks AR, Price D (eds) Chapter 4 in Fire retardant materials, Woodhead Publishing, Cambridge, UK, pp 128–181
6. Miller B, Goswami BC (1971) Effects of constructional factors on the burning rates of textile structures: Part I: Woven thermoplastic fabrics. Text Res J 41:949–954
7. BS 5438:1976 British Standard Methods of test for flammability of textile fabrics when subjected to a small igniting flame applied to the face or bottom edge of vertically oriented specimens
8. ASTM D1230-94, Standard Test Method for Flammability of Apparel Textiles, ASTM International, West Conshohocken, PA, 1994

Design and Development of Smart Sportswear Integrated with Energy Harvesting Device

Yamini Jhanji, S. Khanna, and A. Manocha

1 Introduction

Comfort, although a subjective term, is a crucial property in textile for clothing. Comfort is related to how an individual feels. There are three main aspects for analyzing the comfort of any fabric. The first aspect of comfort is thermal comfort. It is a feeling of pleasure or related to how a hot or cold a person feels. Thermal comfort is associated with changes in many physiological and environmental variables like the activity level of the individuals and clothing properties, such as the fabric insulation values and water vapor permeability. Thermal comfort is mostly quantified using physiological parameters though it is a psychological concept. Tactile sensation is the second aspect of comfort that is related to the interaction of skin with clothing that results from the fabrics in contact with the skin.

The third component of comfort is related to the fit of the garment. A poorly fitted garment, especially too small or too large can interfere mobility and performance, although the impact on comfort may not be as great, it influences the psychological perceptions of the wearer through personal or cultural preferences regarding fit and fashion size trends. Consideration of thermal comfort, sensory skin-feel comfort, comfort due to fit, or the psychological comfort have a considerable impact on the individual physical and cognitive performance.

Clothing is considered to be second skin playing a vital role in maintaining thermal balance with the environment and keep an individual in a comfortable state. It should be the main property of textiles to conserve the heat that body dissipates away, and dissipate heat from body surrounding when the body generates it. Moisture management is crucial for thermo-physiological comfort as it involves controlled

Y. Jhanji (✉) · S. Khanna · A. Manocha
TIT&S, Bhiwani, India
e-mail: yjhanji@gmail.com

V. Midha and A. Mukhopadhyay (eds.), *Recent Trends in Traditional and Technical Textiles*, https://doi.org/10.1007/978-981-15-9995-8_18

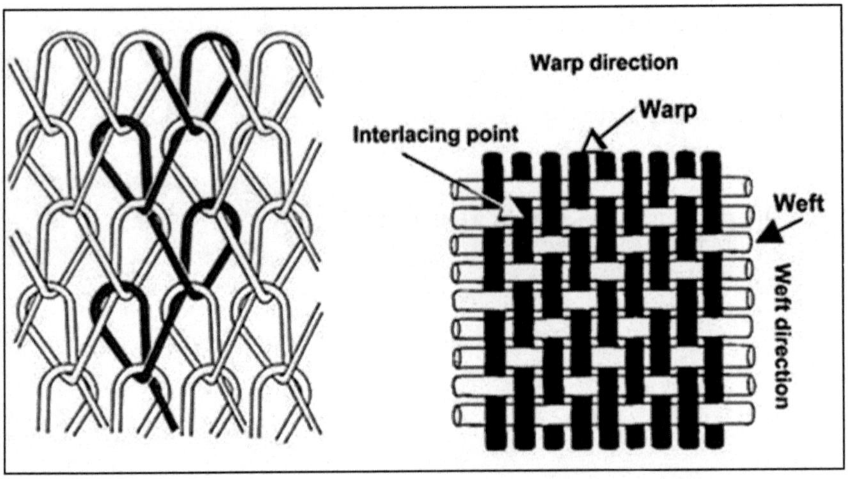

Fig. 1 Knit and woven structure

movement of moisture vapor and liquid perspiration from skin surface to atmosphere through the fabric.

Knitting is a process of fabric manufacturing by interlocking series of loops of one or more yarns. Knitted fabrics are used to produce garments that cover every part of human body, in a wide range of garments types from socks, caps, gloves, and underwear to upper and lower body garments varying from t-shirts to formal jackets. Figure 1 shows a knit and woven structure.

1.1 Requirements of Sportswear

Sportswear are generally involved in sweat generating and rigorous activities like jogging, running, cycling, etc. Therefore sportswear needs to possess the following properties:

- Sportswear garments should be light in weight and dimensionally stable even when wet.
- Outstanding moisture managing properties, which rapidly wicks the moisture away from the body.
- Good perspiration fastness and smart and functional design.
- High electrical conductivity along with superior strength and durability.
- Radiation free

Polyester is the single most common fiber used for sportswear and activewear. Other fibers suitable for activewear are polyamide, polypropylene, acrylics, and elastane. Wool and cotton fibers are still finding applications in leisurewear. Synthetic fibers can either be modified during manufacture, e.g. by producing hollow fibers

Fig. 2 Active and Leisurewear

and fibers with irregular cross section or be optimally blended with natural fibers to improve their thermo-physiological and sensory properties. Synthetic fibers with improved UV resistance and having anti-microbial properties are also commercially available for use in sportswear.

For most sports, the athletes wear a combination of different items of clothing, e.g. sport shoes, pants, and shirts. In some sports, protective gear may need to be worn, such as helmets or American football body armor. Yoga clothing should use fabrics with the good stretchability for easy movement which will likely require the fabric to be of a knitted construction (Fig. 2).

1.2 Wearable Smart Textiles

Smart textiles, also known as smart garments, smart clothing, electronic textiles, or smart fabrics, are fabrics that enable digital components such as a battery and a light (including small computers), and electronics to be embedded in them. Figure 3 shows some wearable smart textiles.

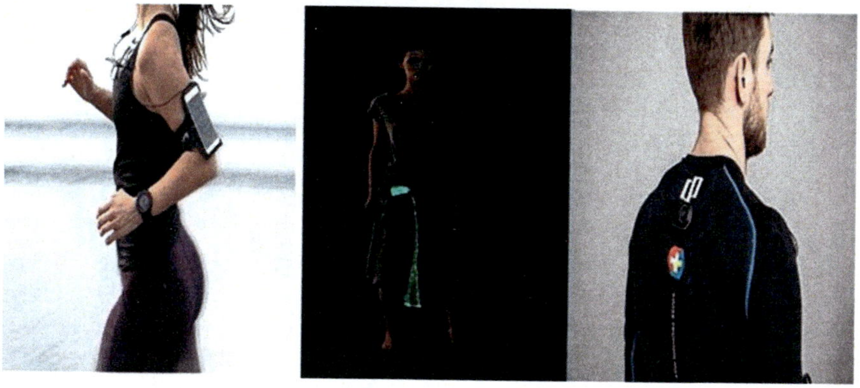

Fig. 3 Wearable smart textiles

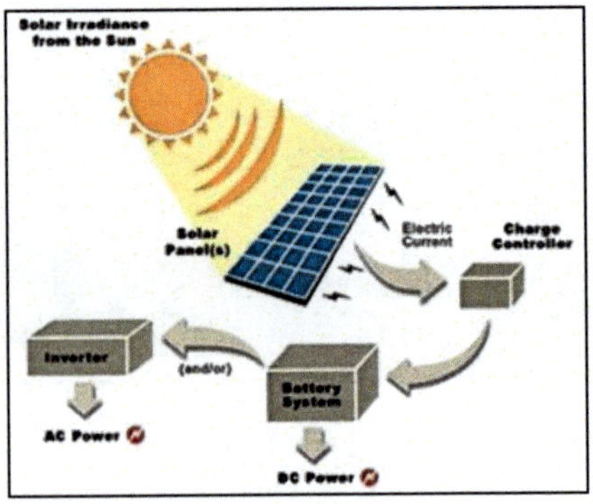

Fig. 4 Working of solar panel

1.3 *Energy Harvesting Devices and Working of Solar Panel*

Energy harvesting, also known as power harvesting or ambient power is the process by which energy is derived from external sources, e.g., solar power, thermal energy, wind energy, salinity gradients, and kinetic energy, also known as ambient energy, captured, and stored for small, wireless autonomous devices, like those used in wearable electronics and wireless sensor networks. Energy harvesters provide a very small amount of power for low-energy electronics. Energy harvesting devices converting ambient energy into electrical energy have attracted much interest in both the military and commercial sectors. The application is in wearable electronics, where energy harvesting devices can power or recharge cellphones, mobile computers, radio communication equipment, etc. Solar panels are those devices which are used to absorb the sun's rays and convert them into electricity or heat. A solar panel is actually a collection of solar (or photovoltaic) cells, which can be used to generate electricity through the photovoltaic effect. These cells are arranged in a grid-like pattern on the surface of solar panels. When the sunlight falls over the solar panel the energy is absorbed in the form of photons and further converted to electric current by the p–n junction diode present in the solar panel. This electric current is then either stored in batteries in the DC power or converted into AC power using the inverters [Fig. 4]. Also, the battery system can be connected to the external USB port in order to transfer the electricity to charge mobiles, torches, power banks etc.

1.4 *Applications of Solar Power in Apparels*

The solar power is now being considered as the future for the generation of electricity due to rapid exhaustion of fossil fuels. Also, solar power has emerged as an economic, sustainable and eco-friendly option to the textile, fashion and apparel industry. From the rooftop solar power plants in the textile industries providing a large amount of electricity to the garments and accessories using the solar power for various decorative, smart and protective features. Figure 5 shows the application area of solar panel in apparels and accessories.

Fig. 5 Application areas of solar panel in apparels and accessories

2 Materials and Methods

2.1 *Materials*

Different knit structures of cotton single jersey and polyester interlock mesh knit structure were used for the study. The fabrics were procured from Vardhman Polytex Ltd., Ludhiana for the study. The details of the fabrics procured have been provided in Table 1.

The solar panel integrated into the designed apparels was procured from Manohar Electronics, Bhiwani. A range of trims and notions like velcro, snap fasteners, and elastic were used.

Table 1 Details of fabric samples and plan for the development of sportswear

Samplecode	Fiber composition	Design aspects	Level of physical activity	End products
SJ_B	PET/C	Wristband	Static/low	Wristband
I_w	100% PET	Zonal garment integrated with solar panel	Dynamic/high	T-shirt, shorts & cap
I_N	100% PET	Zonal garment integrated with solar panel	Dynamic/high	T-shirt, shorts & cap
I_B	PET/C	Zonal garment integrated with solar panel	Dynamic/high	T-shirt, shorts & cap

Note I_w interlock white fabric, I_B interlock black fabric, I_N interlock neon fabric, SJ_B single jersey fabric

The trims and notions were procured from Vinod General Store, Bhiwani for apparel and accessory designing. The trims and notions used in the project were elastics, used in shorts and wristbands for good fit and firmer grip; buttons, for holding the wires of solar panel arrangement; velcros, to firmly hold down the solar panels at their proper position and angle of inclination and also to make the arrangement detachable; piping, used for covering the wires in order to resemble them with the design aspects of the whole garment; interlining, used in the cap for the firmer shape retention.

2.2 Methods

The procured materials were evaluated for their physical and moisture management properties to assess their suitability as sportswear textiles. The prepared fabric samples were evaluated for their fiber content to determine the content or ingredient of provided fabric. The aerial density of samples was determined according to ASTM D-1059. The thickness of fabrics was determined using a fabric thickness gauge. CPI and WPI were measured using the magnifying glass. Moisture management properties of developed knit samples were determined on moisture management tester (MMT) (SDL Atlas, Hong Kong) (AATCC Test method 195-2009) Fig. 6 shows moisture management tester and top and bottom sensors of moisture management tester.

3 Results and Discussion

Different moisture management indices were obtained for the top (next to skin) and bottom (outer layer) for the four test samples under consideration. Analysis of water content versus time curve, fingerprint of moisture management properties are shown in Figs. 7, 8 and 9. It was observed that interlock mesh structures (I_W and I_N) composed of 100% polyester were moisture management fabric as indicated by their fingerprints. Overall moisture management was observed to be highest for I_W and lowest for single jersey structure. I_W fabric was observed to be quick-drying and wicking fabric with lower absorption rate in top (next to skin) layer and higher absorption rate in bottom (outer layer) which is prerequisite for wickable sportswear design. I_N although moisture management fabric has a lower value of AOTI and OMMC as compared to I_W. Moreover, the I_N fabric exhibited lower spreading speed and absorption rate in the bottom layer compared to I_W indicating that fabric would be slow drying and wickless liquid moisture compared to I_W.

I_B although interlock mesh structure exhibited inferior moisture management properties compared to its interlock counterpart with a lower value of OMMC as compared to I_W and I_N.

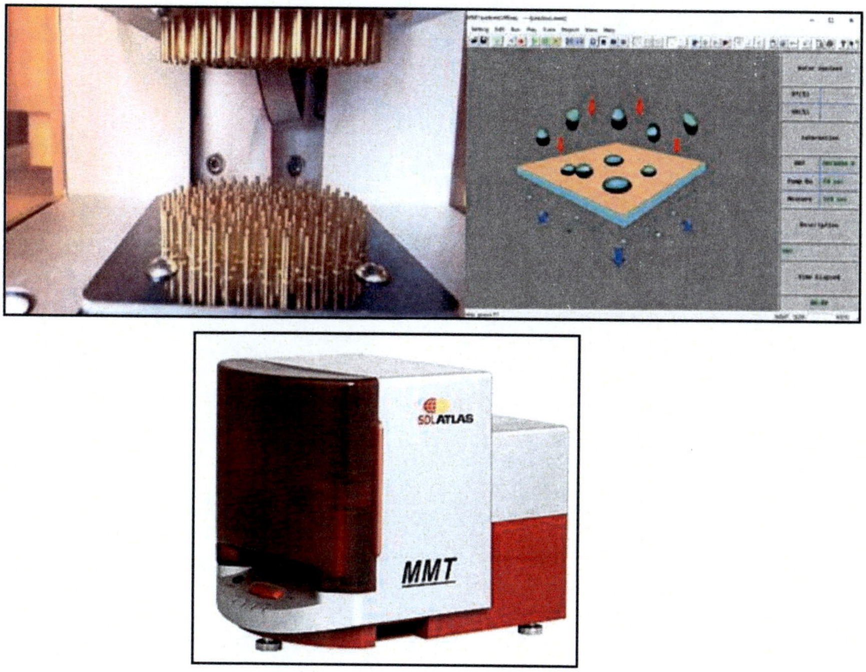

Fig. 6 Top and bottom sensors of moisture management tester

Wetting of the surface is essential to propagate wicking to the outer layer. Hence, such a structure won't be very effective in transferring the liquid to outer layer as wetting of inner layer takes place gradually.

Single jersey fabric (SJ_B) exhibited the lowest OMMC. The results obtained shows that the structure shows higher absorption and spreading speed in next to skin layer (as indicated by AR_t and SS_t), thus making structure suitable for low physical activity or for static conditions in summers.

The results for moisture management properties indicated that Interlock mesh structure (I_W) exhibited excellent moisture management properties. Moreover, the above-stated fabric exhibited the highest one-way transport capability suggesting that structure is capable to keep wearer skin dry with sweat readily transferring from next to the skin to the outer layer.

Results of the study prompted the usage of this knit structure (I_W) for designing sportswear intended for a high level of physical activity.

I_B fabric ranked second compared to I_W as far as moisture management properties are concerned. Therefore, the two mesh structures were predominantly used for designing sportswear.

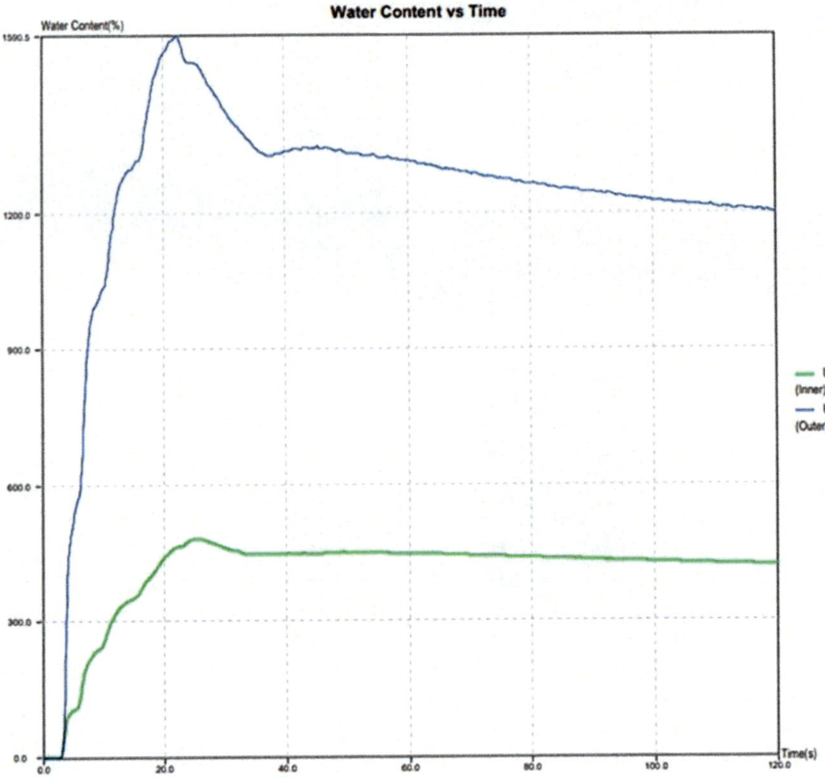

Fig. 7 Water content versus time curves for I_W fabric (I_W—100% P, Interlock mesh)

3.1 *Designing*

Solirt, the solar sports T-shirt complemented with a pair of shorts, baseball cap, and wrist bands were developed using the above-mentioned fabrics that exhibited the moisture management properties suitable for sportswear design.

The design inspiration was taken from sportsperson who give precedence to comfort over the aesthetic appeal. Accordingly, the design was proposed which reflected functionality rather than aesthetic appeal thereby minimal use of surface embellishment of designed and developed apparels and accessories. Furthermore, the design proposed zonal garment with zones to distinguish profusely sweating and low sweating areas of the garment. Mood board and storyboard were designed for the sportswear to be developed. Mood board and storyboard are shown in Fig. 10a, b.

Designing phase was followed by pattern making. Flat pattern making i.e. drafting was employed for the same. A standard size chart was referred for the sizes and

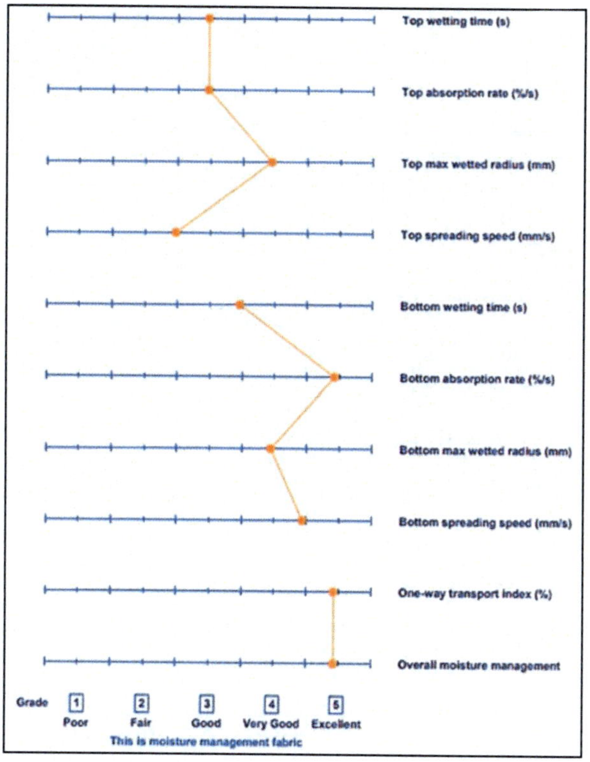

Fig. 8 Fingerprint of moisture management properties of I_W fabric (I_W—100% P, Interlock mesh)

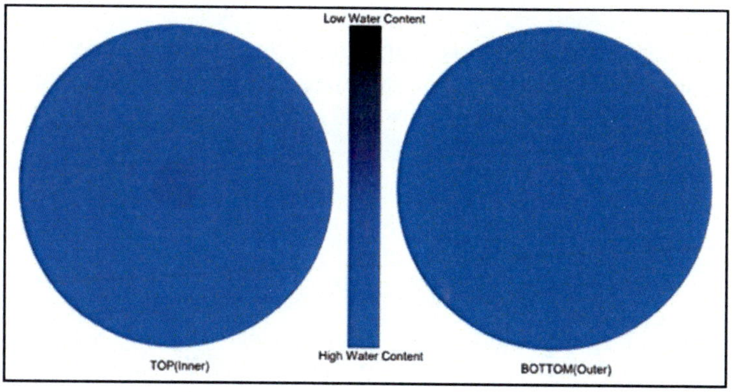

Fig. 9 Water location vs time plot for I_W fabric

Fig. 10 **a** Moodboard, **b** Storyboard

patterns of *L* size were made. Garment assembly designed for sportswear includes t-shirts, shorts, baseball cap and wristbands as shown in Fig. 11.

3.2 *Solar Panel Integration*

It was intended to further enhance the functionality of designed sportswear. Henceforth, solar panels were integrated into **Solirt** t-shirt which served as energy harvesting device and could assist the sportsperson to charge their gadgets, specifically a mobile while on the move or involved in any activity. The Solirt is shown in Fig. 12.

4 Conclusions

- Interlock mesh (100% polyester) structures were the most suitable knit structures for sportswear design and development owing to their excellent liquid moisture transmission properties.

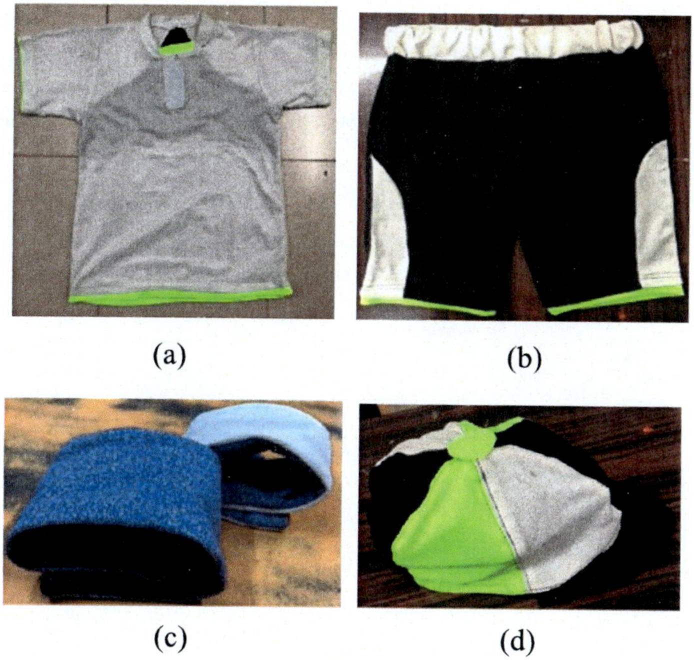

Fig. 11 **a** T-shirt, **b** Shorts, **c** Wrist band, **d** Baseball cap

Fig. 12 Solar panel integrated t-shirt

- Interlock mesh structures exhibited highest one-way transport capability and wicking property compared to their counterparts indicating that such structure would be suitable in transmitting liquid sweat to outer layer effectively, thereby keeping next to skin layer dry.

- The subjective evaluation of designed and developed sportswear by sportsperson involved in dynamic physical activity revealed that sportsperson found the combination of sportswear and accessories functionally as well as aesthically appealing. However, the subjects suggested to reduce the cumbersome solar panel attachments.

Impact of Demonetization on Textile and Apparel Industry

Anju Choudhery, Kiran Choudhery, Varinder Kaur, Parambir Singh Malhi, and Sachin Kumar Godara

1 Introduction

Demonetization of high-value currency means that the particular currency is withdrawn from circulation and replaced with a new currency. The Indian government takes an unprecedented step in the history of the Indian economy by scrapping high-value currency notes of 1000 and 500 with the goal of pushing the Indian economy to new growth heights in the near future, not only to expose black money, stop corruption, counterfeit currency and terrorist financing, but also to boost tax compliance, improve fiscal balance, reduce inflation, encourage cashless transactions and digital India [1, 2]. According to the National Institute of Public Finance and Policy New Delhi "The demonetisation was a big shock to the Indian economy." Demonetization drive has directly or indirectly affected the entire Indian textile value chain in terms of the impact on apparel and fabric demand in the retail sector [3]. More than 80%

A. Choudhery
Department of Education, Guru Nanak Dev University, Amritsar, Punjab, India
e-mail: anjuchoudhery@gmail.com

K. Choudhery (✉)
Department of Economics, University of Rajasthan, Jaipur, Rajasthan, India
e-mail: kiran.economics@gmail.com

V. Kaur · P. S. Malhi · S. K. Godara
Department of Chemistry, Guru Nanak Dev University, Amritsar, Punjab, India
e-mail: varinder_gndu@yahoo.com

P. S. Malhi
e-mail: parambirmalhi@gmail.com

S. K. Godara
e-mail: sachingodara.textile@gndu.ac.in

V. Midha and A. Mukhopadhyay (eds.), *Recent Trends in Traditional and Technical Textiles*, https://doi.org/10.1007/978-981-15-9995-8_19

of the textile industry comes under the MSMEs sector and has been affected significantly. This sector was largely driven by both contractual and daily wage labor and most mobile workers did not have their bank accounts at their place of work [2, 4].

Demonetization is not a new phenomenon; it has been practiced many times in past by several countries such as India (1946 and 1978), Ghana (1982), Nigeria (1984), Myanmar (1987), Soviet Union (1991), Zaire (1993), Australia (1996), European Union (2002), North Korea (2010), Zimbabwe (2010) and Philippines (2015). In previous two demonetization campaign banknotes of 1000 and 10,000 were withdrawn in January 1946 and in January 1978, banknotes of 1000, 5000, and 10,000 were withdrawn [5–8]. Common people were also affected during the previous two demonetization drives but the magnitude of 8 November 2016 drive was mammoth since 87% of the currency was removed out of the system to check black money and terror funding. However, Central Board of Direct Taxes (CBDT) in 2012 has endorsed against demonetization process, bearing in mind that demonetization is not the only solution to remove black money from the system, which was stock primarily in the form of bullion, jewelery, and Benami properties.

The textile industry is one of the Indian economy's oldest industries dating back many centuries. Also today, it is one of India's biggest contributors to total exports. India exports about 50% of total textile and apparel production [6]. It contributes approximately 4% to GDP and 14% to industrial production. Around 40 million people earn a livelihood from the textile and apparel industry. Indian textile manufacturers produce 5% of cloth through the organized market, 20% through the handloom sector, 15% through the knitting sector, and 60% of the fabric through the decentralized power loom industry [9]. India has about 19,42 lakhs of power looms for weaving nearly 19,000 million meters of cloth, and provides more than 7 million people with jobs. The textile industry manufactures a wide range of fabrics from white, dyed, and printed cloth, 100% cotton fabric, blended cotton fabric, wool, silk, linen, regenerated fibers, and other synthetic fibers. The country exports goods worth Rs. 44,000 million to countries such as the USA, France, Germany, India, Hong Kong, Italy, etc. [10]. After agriculture, the clothing and textile industry is the second largest employer in India. The textile and apparel industries played a very important role in generating mass jobs, especially for women, and raised masses of poverty in the world's developing countries [8].

Mostly the reaction of the textile industry to demonetization drive has been mixed one [3]. The organized economy, on the one hand, applauded the push for demonetization as it can rationalize the business transaction mechanism with domestic and overseas traders. The unorganized textile and apparel market, on the other hand, has been severely affected by the demonetization process. The major chunk of the textile industry is the unorganized sector, which consists of handloom, handicrafts, power looms, knitting sector, printing, embroidery, rip and button fixer, fabric cutting, and garment assembly, which are operated on a small scale through traditional tools and methods [10]. This sector requires large-scale labor to produce a wide range of products appropriate for various segments of the market, both in India and around the world. Most of India's clothing and textile units have migrant workers, who are looking for a living for families living in India's rural areas. Textile and apparel manufacturing workers were usually paid in cash as most of them were brought in by contractors and did not have any bank accounts [2, 8]. During the demonetization process, most of these workers were being sent back home because factory owners did not have enough cash to pay salaries. The shortage of new currency badly affected the procurement of raw cotton, yarns and fabric, particularly in the cotton supply chain. Everyone in the textile supply chain has started feeling the heat of demonetization such as farmer, ginning factory, yarn spinning industry, yarn trader, power loom, hosiery industry, fabric traders, garment manufacturer, wholesaler, retailer, and finally consumer. During the demonetization process, debit and credit card use (among consumers and small industries) was significantly increased, and export deals were mostly made via bank transfers. Most of the experts think that the export is a big chunk in the Indian textile industry that segment was not affected by this demonetization move. But the Biswajit Dhar, Professor at the Jawaharlal Nehru University, said that "45% of Indian total export comes from MSME sector, these small exporters were also affected with demonetization drive." Experts overwhelmingly feel that the push for demonetization has mostly helped the organized textile industry, although many unorganized players and companies, operating unethically, may feel heat in the long run to comply [11]. Ministry of Textiles claimed that nearly 5 lakh bank accounts have been opened up to the end of December 2016 in the textile cluster of Surat, Ludhiana, Bhiwandi and Tirupur. The Ministry of Textiles also declared in parliament that 95% of labor wages in jute textile industry of Kolkata were paid through their bank accounts. The most promising result of demonetization process for laborer is that many of the units in the conventional clusters did not properly document provident funds and other employee benefits as most of the transactions were in cash [12]. They would either pay less or on record they won't show them. Therefore, this will not happen again in the future after the worker begins to earn wages in their bank accounts.

2 Objectives

To study the impact of demonetization drive on textile and apparel supply chain.

3 Methodology

In order to study the effect of demonetization drive on the textile supply chain data was taken through various journals, newspaper, e-resources, blogs, reports, and case studies.

4 Effect of Demonetization on Textile and Apparel Supply Chain

The textile and apparel supply chain include various raw material sectors for the manufacture of natural and synthetic textile material, cotton ginning facilities, spinning, wet processing (dyeing/printing/Finishing), weaving/knitting, apparel (other stitched and non-stitched) manufacturer and the distribution channel comprises wholesalers, distributors and a large number of small retailers.

4.1 Impact on Farmers in Cotton Farming

In India, cotton harvesting season begins in September and ends in December. The farmers sell cotton in the month of October to March end. Wardha district is the home to Maharashtra's biggest cotton market. During the cash crunch period, farmers prefer to sell their crop against cash, as this is required for paying cotton picker and transporters. But the commission agents/traders, who buy on behalf of ginners, did not have enough cash to pay. The cotton market was getting only 500–600 cotton-laden vehicles daily, as opposed to the around 1500–1700 normally expected at this time [12]. The low arrivals result in increase in cotton prices to Rs 4800–5000 per quintal from Rs 4100–4200. Some farmers were encashing this opportunity. But the commission agents were paying very small amount around Rs 10,000 in cash or cheque after selling cotton crop of 1–2 lakh rupees approximately. This paltry cash was not sufficient to pay for cotton picking labor and transport cost. The cheque payment adds more inconvenience of standing in long queues, most of the farmers did not have their own bank accounts and only operate on the basis of cash. But in the case of Pathardi market in Maharashtra where the arrival of cotton has reduced by 80%. Some farmers to meet their daily needs were forced to sell their products to local traders even at lower prices because they did not want old notes or any cheque.

4.2 Impact on Ginning Industry

Ginning process is a process to separate the lint fibers from the cottonseed. Ginning is a seasonal business which is generally operated from November to May. During demonetization drive, ginning mills had acute cash shortage and reduced cotton supplies, also the price of cotton increased from Rs. 32,000–35,000 per candy of 355 kg to Rs. 43,000–45,000 per candy in Malegaon [12]. This temporary increase in cotton prices required more cash for paying cotton to suppliers. Most of the ginning mills were also hardly running one shift, instead of three, only to keep labor intact. Mr Patel, a ginning mill owner said that "his factory which was used to run 24 h a day for four months, will now work only for 12 h a day for eight months, since he was receiving cotton from farmers at a slow pace, over a longer period. The labor cost as well as his operational overheads will rise." In Ahmedabad, a ginning unit was shut down because farmers were not ready to accept their payment in the form of cheque. Farmers only accept cash, because they pay farm labour's wages in cash. It was the clear indication that this cashless system cannot work in rural set up [10].

4.3 Impact on Spinning Industry

Spinning is the process for conversion of raw cotton into yarn. The yarn spinning mainly falls within an organized sector, faced the least trouble during demonetization drive, as around 33% of yarn produced is exported to other countries, such as Bangladesh and China [10]. The export business was usually done by bank transfers. Yarn manufacturing companies often supply product directly to yarn traders; they market the yarn to domestic fabric manufacturers. These transactions include intermediate traders who typically buy bulk in cash and sell it in cash to retailers. A few small spinning mills fall under unorganized sectors faced problems in terms of shortage of raw cotton and increase of raw cotton prices, results in price rise of finished yarn. The inventory of finished yarn piled up due to cash crush in small handloom/power loom/hosiery mills, which usually purchase yarn for fabric weaving and knitting. The Indian Texpreneurs' Federation (ITF) has carried out an awareness program in members' of textile mills to educate workers about cashless transactions. Now, most of Tamil Nadu's textile spinning industries deal with cashless transactions to pay their workers' wages and also to buy raw materials [13].

4.4 Impact on Weaving/Power Loom Industry

Weaving is the process for manufacturing of fabric/cloth by the interlacement of warp and weft yarn. Loom is the machine used for manufacturing the fabric materials. Power loom is a motorized machine for weaving cloth/fabric. Around 80% of power loom industry falls under the MSME unorganized sector. Power loom sector was already struggling due to increasing power tariff, competition from cheaper Chinese goods and increasing labor cost. This demonetization drive gave a double blow to already struggling power loom sector and broke down the backbone of the textile industry. According to the New Indian Express, "around 50% of the textile fabric manufacturing units in Sircilla Textile Park, being shut down due to non-availability of yarn during demonetization drive due to lack of cash." The daily production fabric reduced to 1.8 lakh metres from 3.2 lakh meters due to closure of around 700 looms. Due to the nonavailability of yarn for running looms, some of fabric manufacturing unit temporary layoff workforce and told them to look for alternative sources of income temporarily. The city of Banaras is famous for the finest woven sarees made with Zari and Brocade work. The Banarasi sari weave inspired from

Mughal era designs which are based on floral, kalga, bel, pasley, mina work, having intricate borders of gold and silver yarn which are special characteristics of these weaves. Banarasi weaves are like monuments or an intricate piece of artwork which require preservation. Manufacturing of a Banarasi saree on a handloom undergoes many stages such as reeling of silk yarn, degumming, bleaching, dyeing, warping, and weaving on handloom/power loom. Some of these sarees works are worth a few thousands to lakh of rupees. The demand of Banarasi sarees has also reduced significantly due to cash crunch. As customers didn't have money to buy precious saris and the cloth merchants did not have money for the purchase of raw material and paying wages so the handloom/power loom workforce was totally helpless [14]. Due to work stoppages and nonpayment of wages, the majority of the weavers have left jobs. According to social activist, Ehsan Ali, from Banaras, "around 70% of weaving industry has been affected due to demonetization drive." While almost 50% of weavers have opened bank accounts during Jan Dhan Yojana drive but still most of the people who were working in the handloom sector did not get their wages through bank accounts [15].

According to the April 2016 Economic and Political Weekly report, Bhiwandi and Malegoan have about 50% of India's 6.5 million power looms. Around 70% of power looms shut down within one month of demonetization drive. Around 80% of workers in Bhiwandi and Malegaon were migratory in nature and belongs to rural areas of Assam, Bihar, Uttar Pradesh, and West Bengal. Around 75% of the workforce of these two largest textile clusters required a huge amount of cash to pay weekly wages. Laborers partly used the money to meet their daily needs and rest of the money was sent to their families. After demonetization drive, their routine cycle was disturbed [10]. Due to the closure of power looms, they were not having money for food and house rent. Majority of workers were jobless and started returning back to their home town. Asad Farooqi; a labor contactor said that "*Notebandi* ne humko paanch saal peeche fek diya (demonetization threw us five years behind)," Ichalkaranji is well known as the Manchester of Maharashtra having textile processing units as well as cotton-based power looms. Within 15 days of demonetization, daily turnover fell down from Rs 45 crore–Rs 13 crore due to canceling of orders for those who were not accepting payment in demonetized currency [12]. On the other hand business transactions which were usually routed through banks but after demonetization, the cloth merchants were demanding money in cash using the new currency. They were also conveying the caution regarding the cancelation of orders. Bulk orders worth more than 100 crores were canceled and raw material transportation has also been delayed. According to the President of Ichalkaranji Powerloom Weavers Cooperative Association Limited, "The industry was already struggling because of stiff competition. Now, with the government's note-ban bombshell, we were completely crippled [16]." According to I.P. Mahajan, Punjab Textile Manufacturers Association, "power loom sector had more than 50% retrenchment after demonetization in Punjab." Majority of the workforce belonged to Bihar and Uttar Pradesh returned back home after closure of weaving industry [17].

4.5 Impact on Knitting/Hosiery Industry

Knitting is the process for manufacturing of fabric by interloping the yarns. Ludhiana is basically famous for hosiery, woolen knitwear, and garments. It contributes around 80% of the total woolen/acrylic output of the economy and employed more than 400,000 workforces. Ludhiana's hosiery industry is the mother of the industry which has almost 12,000 units catering to the domestic and export market [18]. The unexpected ban of old currency has badly affected the knitting industry; more than 70% of the industries have been shut down temporarily. The woolen and hosiery retail sector has seen a significant decrease in footfall due to cash crunch. According to Vipin Mittal of Kudu Knit Process Pvt. Ltd., "the wholesaler and retailer cancelled or reduced their previous order due to decrease in demand for knitted products, otherwise Ludhiana hosiery was in full swing winter season [17]." According to Ajit Lakra, President, Ludhiana Knitter association, "the knitting unit starts the production for summer collection in November and their manufacturing is at peak in December. However, due to scarcity of cash, most of knitting units were operating at 60–70% of their capacity."

4.6 Impact on Wet Processing (Dyeing/Printing/Finishing) Industry

Amritsar textile industry relies on the winter and wedding season for their sales. In Amritsar alone, more than 30 textile dyeing and printing units have been shut down and the rest of them have cut down their productivity around 25–30%, owing to the cash crunch during demonetization drive. These units were considered as the backbone of around 700–800 textile mills dealing with spinning, weaving, warp-knitting, and embroidery. The dyeing and printing mills closure had a cascading effect on the entire textile supply chain. Textile processer required money to purchase dyes and chemical and to pay labor wages, but they did not have enough cash to sustain. Cash crunch had forced many of them to shut down partially or completely. Most of these units were working one shift or hardly four days a week [17]. Surat is the largest manmade fabric producer in India. It has around 400 textile processing units employing over 3 lakh workers. At Pandesara, Surat, most of textile processing units are usually operated by the contract based worker. The factory owners faced problems in paying salaries. Around 40% layoffs in textile processing units was estimated. According to the Federation of South Gujarat Textile Processing Association president named Jitu Vakhariya, "they were surviving by reducing production and running units hardly 3–4 days a week with old job works." Some of the processing units were supported by labor contractors due to their long-term relationship with a promise to pay them when things get better. Wages had been also gone down during this period [13].

4.7 Impact on Garment Manufacturing Industry

Tirupur is the biggest knitwear cluster of garment export in India. Tirupur has more than 5000 garment manufacturing and job work units [4]. Tirupur garments industry always have a shortage of skilled labor throughout the year but within one month of demonetization drive instead of "Employees Required" signboards replaced with "To Let" due to shut down of big chunk of MSME's garment houses [19]. Tirupur cluster alone employs around 5–6 lakhs workforce directly or indirectly and gives the business of around 40 K crore. An estimated export turnover was 30 K crore till 2017 March was shortfall by 4 K crore may be due to this demonetization drive. Tirupur apparel industry mainly haves three types of units: 100% export-oriented, buying houses who subcontract work from many small and bigger units depending upon their requirement and domestic garments houses which caters to the domestic Indian market. A major chunk of garment export business deals with cashless transaction worth of approximately 22 K crore and least affected segment with demonetization drive. However, cash crunch during demonetization drive has significantly affected small exporter and domestic garment houses, subcontract work of buying houses and job worker; they deal with cash transaction for paying wages, buying fabric, garment accessories, trims and transport charges. Badly affected people in the entire value chain were small job workers who work from home or small shop dealing with embroidery, buttoning, small servicing, printing, and packing. A labourer who can earn Rs 500–600 daily was standing in queue for exchange of older currency notes. According to the General Secretary of Tirupur Exporter Association, 25% of subcontracts and job work commissioned by exporter affected due to cash crunch. Delay in garment export consignment put an additional burden of 20% penalty in the form of costly airlifting of cargo. A resolution was passed by the South Indian Shirt Manufacturer Association to reduce the workforce from 15 workers per shift to 6 per shifts. Most of the garment manufacturers were layoff non-salaried workers and contract workers. Around 70% of the garment industry in Tirupur falls in MSME sector and badly affected by demonetization. Many garment manufacturer who might have unaccounted cash or black money were paying 3–4 months salaries in advance in old currency notes of denomination of Rs. 500 and 1000. Laborers' were finding it tough to exchange cash in long queues for meeting daily needs instead of working [4, 20–23]. Demonetization drive was appreciated by Tirupur garment exporter because most of the unorganized unit works on unaccounted cash transaction of around 7 K crore. These garment houses were not paying any taxes. If these garment units come under tax net there will be the level playing field for all [24].

4.8 Impact on Fabric Retailing

More than 90% of the transaction is done by cash in Indian fabric retailing sector. Consumers pay cash to retailers who pay wholesalers or cloth merchants most of the time on a weekly, bimonthly or 90-day basis, as this is convenient. Wholesalers or fabric merchants place large orders with fabric manufacturing factories and pay through cheques or bank transfers; they didn't do that due to cash shortage during the demonetization process, followed by a significant reduction in retail spending. Most retailers and wholesalers have avoided a number of taxes by cash transactions: the volume of trade is very difficult to record with many transactions off the records. Nowadays, digital transaction or internet banking dominates the bulk purchases of fabric that traders directly made to fabric manufacturing factories in Ahmedabad, Bhiwandi, Coimbatore, Ludhiana, and Surat [10]. The historic Mulji Jetha market (180-years old bazaar) in Mumbai is the largest wholesale textile cloth market in Asia. According to the market chairman, "during peak wedding season from November–January its turnover exceeds Rs 100 crore/day." But within a week of demonetization exercise, suiting and shirting fabric, designer saris, churidar salwar, bed-sheets, lungis, towels, and curtains were lying unsold in the form of massive stacks of fabric, reported around 70% fall in sales. Almost 70% of the market caters to wholesalers, hardly 30% gets retail customers. Many consumers stopped buying luxury apparel while low-end item purchases were not much affected [10]. The wholesale trader of this market sells fabric to small apparel industries, small wholesaler, distributors, and retailers across the country, including in Andhra Pradesh, Bihar, Delhi, Haryana, Kerala, Kolkata, Madhya Pradesh, Punjab, Tamil Nadu, Uttar Pradesh, etc. The wholesale cloth and yarn merchant association in Nagpur welcomed the demonetisation decision, but around 80% falls in cloth sales as most of the cloth purchases were done in cash during the peak marriage and festival season. A similar situation was reported in Ahmedabad's new cloth market, where clothing trade had reduced by 80%. Most of embroidery job workers, who were working for wholesale merchants returned, back home after demonetization [9]. As sales fell, cash dried up except for exports, export revenues receive through a bank account so he couldn't pay wages because most of the workers had no bank account and they usually pay wages in cash. Besides, in small district towns of Vidarbha such as Bhandara, Gondia, Gadchiroli, and Chandrapur, retailers chiefly depend on the farming community for their sales [16]. Farmers were struggling to meet their daily needs, it would not be to see cloth markets, too, coming to a standstill.

4.9 Impact on Garment Retailing

Majority of Indian population is used almost 80% cash for buying essentials, including clothes. Most of famous garments brand and retailer reported a dip in 40–60% sales in the first few weeks of demonetization [8]. According to ICRA's assessment, the effect of demonetisation was most serious for winter wear retailers and apparel manufacturers dealing with the domestic market, which accounted for 60–70% of their annual sales during October to February, span [25]. The garment manufacturer delivers deliver by September–October but the wholesaler's sales pressure through the retailer during the peak season will indirectly affect manufacturers.

Lesser sales increase liquidity pressures on retailer and result delayed payment to the manufacturer [26]. Neighboring states like Haryana, Himachal Pradesh, Jammu and Kashmir, Rajasthan, and Uttarakhand shrunk Ludhiana's hosiery apparel demands after demonetization drive. During the month of December, retailers of woolen clothing were forced to offer large discounts of 20–50% on fresh arrivals in a bid to clear the built-up inventory. Usually, discounts were offered after mid-January or February [27]. Domestic Branded retailers were forced to introduce various promotional schemes and attractive discounts 4–5 weeks ahead of their schedule to counter this demonetization effect. Even International retailers with fast-fashion chains like *Zara* and H&M were forced to offer preseasonal sales to clear their stock quickly [28]. Pepe Jeans reported a 49% reduction in sales on the very next day of the demonetization and gave an additional 5% discount on the use of credit/debit cards to minimize their loss. This technique helped them to cover up to to 25% of losses. Indian's largest shirting and suiting brand Raymond also recorded a 30% decline in demand after the demonetization process, and it came at a time when shopping for the wedding season was at its peak. According to the Turtle company, "plastic money and other payment methods have held the fashion retail alive in bigger cities." Within a week of demonetization process, e-commerce giants Flipkart, Amazon, Snapdeal, and Paytm announced the upswing in their market, most online retailers stopped picking up the COD (cash on delivery) in old currency notes. COD contributes around 70% sales of e-commerce business and after halting cash on delivery, business of Flipkart, Amazon, and Snapdeal also reduced significantly. Definitely a rapid growth in credit/debit card/online/e-wallet transaction was triggered by demonetization. But the Indian textile business' bigger pie depends on cash transactions, and the effects of demonetization seen in the form of a dip in the profits. The effect of demonetization was felt most by small traders and the unorganized retail segment that relies heavily on cash for trade and sales [2]. The decline in the share of clothing sales for an organized player was less compared to an unorganized player. The ethnic wear sector had been badly hit due to the peak wedding season during demonetization drive.

5 Finding

- Cotton arrivals in the market were delayed because of the cash crunch. Many farmers hold their inventories and results in temporary price rises.
- Around 50–70% of the Micro Small and Medium Enterprises textile industries have either temporary shut down or reduce their production and even some mills barely run single shift to hold their trustworthy old workers.
- MSME sector was most affected but organized big corporate houses like Raymond, Vardhman, Arvind mills, Gokaldas Exports, Madura Garments and Shahi Exports were least affected during demonetization drive.
- Textile mills told their workers to temporarily search for alternative sources of income and redundancies were rampant in the textile clusters.

- Layoff of workers was due to nonavailability of raw materials, unable to pay wages due to cash crunch, cancelation of big orders, and no demand for clothing and apparel in the market after demonetization.
- Raw material shortage in the cotton supply chain: shortage of raw cotton for ginning mills and yarn manufacturers. Yarn shortage for knitting (hosiery) and weaving (power loom and handloom) units. Fabric shortage for the textile processor and garment manufacturer. Dyes and chemical shortage for the wet processing unit.
- Most of the textile sector workers were migrants from Andhra Pradesh, Bihar, Maharashtra, Rajasthan, Orissa, Uttar Pradesh, etc.
- The wages in some textile clusters were also reduced after the demonetization drive.
- Organized textile sector appreciated the demonetization move. They thought demonetization drive will rationalize the transaction system resulting in more transparent business.
- Demonetization drive brought the unorganized textile sector under the tax net after transparent business transaction through bank accounts and there will be the level playing field for everyone.
- In old notes, some of the textile mills paid wages 2–3 months in advance to the workers. This move made an exchange of old notes for workers standing in long queues. Yarn spinning sector was least affected among the textile value chain, farmer and retailer were the most affected ones.
- The export was less affected during demonetization as compared to the domestic textile market.
- Around 50–80% of the labors did not have a bank account, who were working in the unorganized sector or under a contractor in different textile clusters of India. Large numbers of mills also opened bank accounts for the workers during the demonetization drive with government initiatives, and charged their salary in the accounts.
- Winter wear and ethnic wear were the most affected segment among apparels because the domestic market witnessed about 60–70% of their annual sales during the November–February period during the festival and marriage season.
- E-commerce companies like Flipkart, Amazon, Snapdeal, and Paytm had a significant surge in their business during demonetization drive, but most of these companies stopped the cash on delivery within two weeks.
- There were two extreme effects of demonetization effects, one side there was a temporary price rise of raw cotton because the farmer's didn't prefer selling due to nonavailability of cash. On the other side, garment retailers were forced to give discounts to cop up with buildup inventory.

6 Conclusion

Demonetization drive has been largely shown a mixed response among the textile industry. On the one hand, the demonetization drive was appreciated by organized textile sector as it would restructure the transaction system and made business more transparent and level playing field with unorganized industrial/retailer which were not paying taxes, but on the other hand unorganized textile sector was crumpled due to cash crush there is no demand, shut down of looms, retrenchment of worker, build up yarn, fabric and garment inventory and on the other side shortage of cotton, yarn, fabric, dyes, and chemicals for small scale units. The demonetization drive has also affected the purchase of raw cotton, new yarns and fabric and deferred the payment of wages in the unorganized textile sector. Everybody in the textile value chain was also affected during demonetization drive from cotton farmer to garment retailer. Loss of production, job retrenchment, and temporary price become prevalent in major textile clusters of India. The export of textile and readymade garments was also affected during demonetization drive but the extent was less as compared to domestic fall in demands during peak wedding and festival season. Increase in awareness about digital mode payment, large number worker opened a bank account in the major textile cluster of India like Ludhiana, Tirupur, Surat, Bhiwandi, Malegaon, and Ahmedabad. There has been a significant rise in digital mode of payment post demonetization, although their base is still small.

References

1. Parashuramulu B (2017) Impact of demonetization on Indian economy-merits and demerits. Zenith Inter. J Bus Eco Mgmt Res 7(9):66–76 (2017)
2. Anon (2017) National institute of public finance and policy. Demonetisation: impact on the economy. https://www.nipfp.org.in/media/medialibrary/2016/11/WP_2016_182.pdf
3. Sharma SP, Areesha (2017) Impact of demonetization on economy, businesses and people suggestive measures for remonetisation. https://phdcci.in/image/data/Research%20Bureau-2014/Economic%20Developments/paper/Study%20on%20Impact.pdf
4. Gulati GS (2016) Demonetization and its impact on Indian garment industry. https://www.magzter.com/article/Fashion/Business-Of-Fashion/Demonetisation-and-its-impact-on-Indian-Garment-Industry
5. Arun S (2016) Demonetisation may hit labour-intensive export sector output. https://www.thehindu.com/business/Industry/Demonetisation-may-hit-labour-intensive-export-sector-output/article16671262.ece
6. Bhadouriya D (2017) Demonitization-impact on various sectors in India. https://www.slideshare.net/deepakbhadouriya1/demonitization-impact-on-various-sectors-in-india
7. Kumar S, Bumra S (2017) Demonetization in India: pre and post effect on Indian industry. Inter J Sci Eng Res 8(7):266–274
8. Shawl JA (2017) Demonetization and its impact on the indian economy. J Res S Sci 7(11):456–473
9. Bhoje G (2017) Victims of new economy and crisis in power loom sector bhiwandi nizampur municipal corporation. Inter J Res Eng IT S Sci **7**(9), 43–47 (2017)

10. Waghmare A (2017) Demonetisation: how India's textile industry is tottering after note ban. https://www.business-standard.com/article/economy-policy/demonetisation-how-india-s-textile-industry-is-tottering-after-note-ban-117012300160_1.htm
11. Anon (2016) Nirmala Sitharaman hopeful of positive export growth next year. https://economictimes.indiatimes.com/articleshow/56054481.cms?from=mdr&utm_source=contentofinterest&utm_medium=text&utm_campaign=cppst
12. Paigude S (2016) In Manchester of Maharashtra, note ban silences textile industry. https://economictimes.indiatimes.com/news/economy/foreign-trade/nirmala-sitharaman-hopeful-of-positive-export-growth-next-year/articleshow/56054481.cms?from=mdr
13. Neelam (2017) Impact of demonitisation on Indian textile industry. Inter J Adv Res Dev 2(5):224–226 (2017)
14. Kahlon AK (2017) Effect of demonitisation on weavers of Varnasi a conceptual overview. Inter J Sci Tech Mgmt 6(1):363–366
15. Singh P (2016) Demonetisation adds to weavers' woes. https://businessworld.in/article/Demonetisation-Adds-To-Weavers-Woes/24-11-2016-108772/
16. Sagar A (2016) Impact of demonetisation on textile industry of Maharashtra. Inter J Mgmt App Sci 2(11):169–173
17. Vasudeva V (2016) Punjab's famed textile industry finds itself on pins and needles. https://www.thehindu.com/news/national/Punjab's-famed-textile-industry-finds-itself-on-pins-and-needles/article16901129.ece
18. Ani (2016) Ludhiana hosiery industry badly hit by demonetization (2016) https://www.business-standard.com/article/news-ani/ludhiana-hosiery-industry-badly-hit-by-demonetization-116112100092_1.html
19. Buvaneswari G (2016) Impact of global economic crisis on Tirupur textile industry. https://www.indiantextilejournal.com/articles/fadetails.asp?id=4802
20. Nick (2017) Demonetisation and its effect on the Indian textile industry. https://www.bizvibe.com/blog/demonetization-and-its-effect-on-the-indian-textile-industry/
21. PTI (2016) Demonetisation may hit labour-intensive export sector output. https://www.thehindu.com/business/Industry/Demonetisation-may-hit-labour-intensive-export-sector-output/article16671262.ece
22. Thangavelu D (2016) Textile hub Tirupur still reeling from demonetisation blow. https://www.livemint.com/Industry/g0n3yjjxdwoa9rydh6ufql/textile-hub-tirupur-still-reeling-from-demonetisation-blow.html
23. Allirajan, Jain A (2016) Tirupur garment industry tides over demonetisation crisis. https://timesofindia.indiatimes.com/city/coimbatore/Tirupur-garment-industry-tides-over-demonetisation-crisis/articleshow/55452910.cms
24. Srividya PV (2017) The warp and woof of Demonetization. https://www.thehindu.com/news/national/tamil-nadu/the-warp-and-woof-of-demonetisation/article16667604.ece1
25. Burea FB (2016) Demonetisation impact: Textile sector to be impacted in near-term. https://www.financialexpress.com/market/commodities/demonetisation-impact-textile-sector-to-be-impacted-in-near-term-says-icra/461853/
26. Ghosal S (2016) Demonetization hits disposable incomes and sales of textile goods. https://economictimes.indiatimes.com/industry/cons-products/garments-/-textiles/demonetisation-hits-disposable-incomes-and-sales-of-textile-goods/articleshow/55727890.cms
27. PTI (2016) Demonetisation: Punjab's hosiery sector facing demand slump. https://indianexpress.com/article/india/demonetisation-punjabs-hosiery-sector-facing-demand-slump/
28. Ftrend (2017) Demonetisation and its impact on fashion Industry—India. https://f-trend.com/en-in/blog/demonetisation-and-its-impact-fashion-industry-india

Studies on the Water Retention Properties of Coir Needle Punched Nonwovens

J. C. Sakthivel and L. Sivashankar

1 Introduction

Agricultural sector is one of the sectors which face many threats in the current scenario due to various reasons such as climate changes, water scarcity and modernization. Among this water scarcity and water management are the serious problems. Textile is one of the suitable sectors which can offer a better solution to this. The unique textile properties like large surface area, light weight, higher water retention properties, eco friendly and biocompatibility has made them a better alternative; i.e. water retention nonwoven bed is the better solution to this. Agrotech is one of the major areas in the group of technical textiles. Textile structures in various forms are used in shade house, green house and also in open fields to control environmental factors like temperature, water and humidity. Natural biodegradable textile fibers are grown as agricultural plants and are commonly used for the production of ropes, carpet backings, hand bags, etc. They are classified into three groups like bast fibers (flax, hemp, jute and kenaf), leaf fibers (sisal, pineapple and banana) and seed or fruit fibers (cotton and coir).

Coir is a biodegradable organic fiber and hardest among other natural fibers, shows more advantageous in different application for agricultural textiles. The fiber is hygroscope with moisture content of 10–12% at 65% relative humidity. Coir is having a very high potentiality in agro textile application. Its moisture retention capability and high wet strength has been excellent and the characteristic has been made use extensively in agro textile applications [1]. In polypropylene nonwovens, the soil moisture retention functions were influenced both by pore size and geo textile fiber-water contact angle, and the relative influence of each factor varies

J. C. Sakthivel (✉) · L. Sivashankar
Department of Textile Technology, PSG College of Technology, Coimbatore, Tamil Nadu, India
e-mail: jcs.textile@psgtech.ac.in

V. Midha and A. Mukhopadhyay (eds.), *Recent Trends in Traditional and Technical Textiles*, https://doi.org/10.1007/978-981-15-9995-8_20

among products [2]. The higher moisture absorption properties of the drum stick fibers create the possibilities in applications in geo and agriculture textiles [3].

Agro-tech possess various desirable properties such as protection from insects, light(ultra violet rays) or hail, lightweight, biodegradability, resistance to microorganisms, and high potential to retain water. Weather resistance, it must work effectively in cold as well as hot climatic conditions, resistance to microorganisms-it must resistant to microorganism to protect the living being, stable construction—the construction must be such that it must be stable for any application, lightweight—the weight of the fabric should be such that it will bare by the plant, withstand solar radiation, withstand ultraviolet radiations, high potential to retain water and protection property [4]. One of the requirements to be met by agricultural nonwovens is high potential to retain water so seeds can germinate and plants can grow. This is achieved by means of fibre materials which allow taking in much water and by filing in super absorbers [5].

In this research work, an attempt has been made to study the physical properties and water holding capacity of coir needle punched nonwoven and also study the utilization of water retention nonwoven bed for assistance in plant growth. Needle punched nonwoven samples was developed with coir/cotton and coir/kenaf fibres with 90/10 blend ratio. The physical properties and water holding capacity of the coir/cotton and coir/kenaf needle punched nonwoven was studied. The effect of coir/cotton and coir/kenaf fibres needle punched nonwoven water retention bed on the plant growth also studied.

2 Experimental

The natural fibre like coir, MCU5 cotton and kenaf fibers was selected for the study. The properties of the cotton fibre; 2.5% span length is 32.5 mm, Micronaire is 3.8 and tenacity of the fibre is 24 g/tex. The properties of the kenaf fibers; staple length is 40 mm, denier is 1.2 and tensile strength of the fibre is 5.7 gpd. The coir, cotton and kenaf fibre was opened and cleaned by using miniature carding machine. The cut raw coir fibre and cotton and kenaf fibre was mixed manually based on the required blend proportions.

The carded webs of coir, kenaf and cotton with blend proportions 90:10 was produced with TRYTEX miniature carding machine. The 10% of cotton and kenaf fibre was used to improve the stability of the needle punched nonwovens and to study its effect on water retention properties of the fabrics. The coir/cotton and coir/kenaf needle punched nonwovens was developed using DILO needle punched nonwoven machine. The thickness, areal density, air permeability and water holding capacity of the developed needle punched nonwoven samples were tested. The needle punched nonwoven samples was tested for its water holding capacity.

To test water retention capacity of nonwovens, first a tray of dimension 11 × 8 × 4 inches was taken and a hole of 5 mm is drilled at the right bottom corner of the tray. Then the sample size of 8 × 8 inches was taken and placed inside the tray.

150 ml of water was sprayed randomly over the nonwoven bed using a water sprayer. Then the amount of water drained was noted at periodic intervals till the bed reached its saturation level. Their corresponding value was used. The same procedure was repeated to test the water holding capacity of soil alone. Soil was filled in the tray in accordance with thickness of the sample. The same procedure was followed with soil filled in the tray for 15 mm and the sample placed above it. The amount of water drained was noted and their results was tabulated.

3 Results and Discussion

Table 1 shows the physical properties of the needle punched nonwoven samples was developed with coir/cotton and coir/kenaf fibres with 90/10 blend ratio.

The thickness and areal density of the coir/kenaf needle punched nonwoven was quite higher compared to that of coir/cotton needle punched nonwoven. It may be due to the, coarseness of the kenaf fibres compared to that of cotton. The air permeability of the coir/kenaf needle punched nonwoven was lower compared to that of coir/cotton needle punched nonwoven, the coir/kenaf sample doesn't allow much air to pass through the samples, coir/kenaf was denser than the other samples.

3.1 Effect of Time on the Water Retention Capacity of Coir/Cotton and Coir/Kenaf Needlepunched Nonwovens

The water retention capacity of the coir/cotton and coir/kenaf was tested, in addition to that water retention capacity of soil, water retention capacity of soil with needle punched nonwoven also tested. Table 2 shows the amount of water drained with respect to time of the coir/cotton and coir/kenaf needle punched nonwoven.

It was clear that coir/cotton needle punched nonwoven have drained in the given period of time was low or water retention property of the coir/cotton needle punched nonwoven was higher. The results show that, the water holding capacity % of the

Table 1 The physical properties of the coir/cotton and coir/kenaf fibres (90/10)

S. No	Properties	Coir/cotton	Coir/kenaf
1	Thickness, mm	14.72	15.13
2	Areal density, g/m^2	977	1057
3	Air permeability, cm^3/cm^2/s	352	242
4	Packing density	0.0474	0.0499
5	Fabric porosity, %	95.26	95.01

Table 2 The amount of water drained with respect to time of the coir/cotton and coir/kenaf needle punched nonwoven

Time (min)	5	10	15	20	25	30	35	40
Coir/cotton	75	83	90	96	97	97	97	97
Coir/kenaf	40	98	108	113	113	113	113	113

Table 3 The amount of water drained with respect to time of soil with different thickness level

Time (min)	5	10	15	20	25	30	35	40
13 mm soil height	5	11	15	16	18	18	18	18
14.5 mm soil height	2	9	12	15	16	17	17	17
15.5 mm soil height	3	8	10	12	15	15	15	15

coir/cotton needle punched nonwoven is 35.34% and the coir/kenaf needle punched nonwoven are 22%.

3.2 *Effect of Time on the Water Retention Capacity of Soil with Different Thickness Level*

The water retention capacity of soil alone was tested. Soil was filled in the tray with 13, 14.5- and 15.5-mm height and then water was sprayed evenly over the soil surface. Table 3 shows the amount of water drained with respect to time of the coir/cotton and coir/kenaf needle punched nonwoven.

It was observed that, there is no significant difference in water holding capacity of soil with different height. The result shows that, the water holding capacity % of the soil with 13 mm height is 88% and the soil with 15 mm height is 90%.

3.3 *Effect of Time on the Water Retention Capacity of Coir/Cotton and Coir/Kenaf Needlepunched Nonwovens with Soil*

The combined water retention capacity of coir/cotton and coir/kenaf needle punched nonwovens with soil height of 15 mm was tested.

From Fig. 1 it was clear that only 10–13 ml of water drained from the sample and the remaining water has been retained by both needle punched nonwovens and soil. It was observed that, water starts drained out after 25 min only. This indicates that it has reached the soil and from the soil it took some more time to drain out.

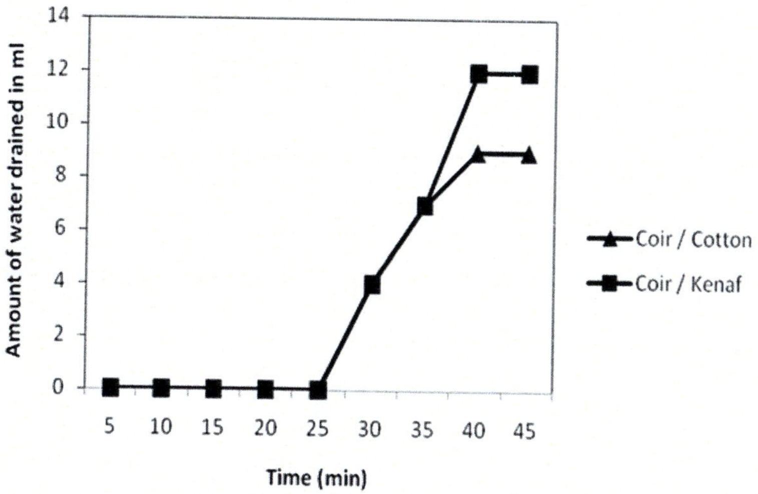

Fig. 1 Comparison of amount of water drained from coir/cotton and coir/kenaf needle punched nonwovens with soil height of 15 mm

3.4 Effect of Coir/Cotton and Coir/Kenaf Needlepunched Nonwovens with Soil on Plant Growth

The assistance of coir/cotton and coir/kenaf needle punched nonwovens with soil for plant growth was conducted. The growth of the plant was measured with respect to days. It was observed that, the height of the plant was very low for the plant without nonwoven bed. So, it was clear that the coir/cotton and coir/kenaf needlepunched nonwoven assists the plant growth by maintaining the microclimate between the top layer of the soil and the nonwoven bed.

The plant with coir/cotton needle punched nonwoven shows the better growth compared to that of other sample. It was observed that, the height of the plant with coir/cotton needle punched nonwoven bed is 19.6 cm and coir/kenaf needle punched nonwoven bed is 17.8 cm as shown in Table 4.

4 Conclusions

The needle punched nonwoven beds was produced with coir/cotton and coir/kenaf using needle punching technology. It was observed that, needle punched nonwovens produced with 90% coir/10% cotton shows better results in terms of water retention capacity and higher plant growth. The major outcome of this research work was, coir-based needle punched nonwovens creates better microclimate, water retention bed reduces the water consumption and support the better plant growth.

Table 4 Effect of coir/cotton and coir/kenaf needlepunched nonwovens with soil on plant growth (cm)

Days	5	6	7	8	9	10	11	12	13	14	15	16	17	18	19	20
Coir/cotton	0	1.3	2.65	3.9	5.3	6.6	7.9	9.25	10.5	11.8	13.2	14.5	15.85	17.2	18.5	19.6
Coir/kenaf	0	1.2	2.6	3.6	4.8	6.1	7.2	8.4	9.5	10.7	12	13.2	14.5	15.6	16.7	17.8

References

1. Sudhakaran Pillai M, Vasudev R (2001) Applications of coir in agriculture textiles. In: International seminar on technical textiles, Mumbai, June, pp 1–5
2. Stormont JC, Henry KS, Evans TM (1997) Water retention functions of four nonwoven polypropylene geo textiles. Geosynth Int 4(6):661–672. ISSN 1072-6349
3. Sakthivel JC, Vishnu SA (2019) Studies on drumstick fibre for textile applications. Manmade Text India 47(6):185–187. ISSN 0377 7537
4. Surya R, Prabhu V, Sakthivel JC (2019) Agro-textiles: requirements and applications. Indian Text J 120–123. ISSN 0019 6436
5. Albrechet W, Fuchs H, Kittelmann W (2003) Nonwoven fabrics. Wiley-VCH, Germany. ISBN 3-527-30406-1